THE BOOK ON PROJECT MANAGEMENT

How to be Extremely Efficient and Remain Profitable While Developing Any Project

GUSTAVO A. VALENZUELA
WWW.THEBOOKONPM.COM

The Book on Project Management
How To Be Extremely Efficient and
Remain Profitable While Developing any Project
www.thebookonpm.com

Copyright © 2018 Gustavo A. Valenzuela

ISBN-13: 978-1986879231

All rights reserved. No portion of this book may be reproduced mechanically, electronically, or by any other means, including photocopying, without permission of the publisher and the author except in the case of brief quotations embodied in critical articles and reviews. It is illegal to copy this book, post it to a website, or distribute it by any other means without permission from the publisher and the author.

Limits of Liability and Disclaimer of Warranty
The author and publisher shall not be liable for your misuse of the enclosed material. This book is strictly for informational and educational purposes only.

Legal / Financial / Investment Disclaimer
This book and the accompanying website are for information and illustrative purposes only and do not purport to show actual results. It is not, and should not be regarded as investment advice or as a recommendation regarding any particular security or course of action. Opinions expressed herein are only current opinions of the author, as of the date appearing in this material only and are subject to change without notice. Reasonable people may disagree about the opinions expressed herein. In the event that any of the assumptions used herein do not prove to be true, results are likely to vary substantially. All investments entail risks. There is no guarantee that investment strategies will achieve the desired results under all market conditions and each investor should evaluate his ability to invest for the long-term especially during periods of a market downturn. No representation is being made that any account, product, or strategy will or is likely to achieve profits, losses, or results similar to those discussed if any. This information is provided with the understanding that with respect to the material provided herein, you will make your own independent decision with respect to any course of action in connection herewith and as to whether such course of action is appropriate or proper based on your own judgment, and that you are capable of understanding and assessing the merits of a course of action. The author, publisher and accompanying website do not purport to and do not, in any fashion, provide tax, accounting, actuarial, recordkeeping, legal, broker/dealer or any related services. You may not rely on the statements contained herein. The author, publisher and accompanying website shall not have any liability for any damages of any kind whatsoever relating to this material. You should consult your advisors with respect to these areas. By accepting this material, you acknowledge, understand and accept this foregoing.

Publisher
10-10-10 Publishing
Markham, ON Canada

Printed in Canada and the United States of America

Table of Contents

Dedication Page — vii
Foreword — ix
Acknowledgements — xi

Chapter 1 What is Project Management? — 1
Official Definition for Project Management — 1
Project Management Styles — 3
LOPPM *(Lack of Proper Project Management)* — 9
Who Runs Project Management? — 12
What is and What IS NOT Project Management — 14
A Robust Definition for Project Management — 15

Chapter 2 Why Project Management — 17
Projects Developed by Inexperienced Staff — 17
Effects of Having an Experienced Project Manager — 19
Losing Money — 21
Losing Control — 22
Pretty Schedules/ Unwanted Results — 24
Accountability of Stakeholders — 26

Chapter 3 Phases in Project Management — 29
How Many Phases are in PM? — 29
Going from A to B — 31
The Crucial Tasks Between A and B — 32
When to Engage a Project Manager — 34
PM Duties During Project Development — 35
Efficiency Definition through PM's Lens — 36

Chapter 4 How to Get the "A-Team"   39
What Matters Most During the Selection Process   39
Procurement Beyond Legal Requirements   40
Statement of Qualifications   41
The Right Interview Questions   43
A Conclusive Selection   45
Let it Flow; It's the Right Team!   47

Chapter 5 Efficient Ways to Get Your Project Done   49
What is "Efficient?"   49
Organization Based on Efficiency and Flow   50
The Knowledgeable Team Effect   52
The Experienced Team Effect   52
Recognize the Actual Status of Your Project   53
Who Was Before and Who Comes After.

Chapter 6 The Genius Way to Organize Your Team   57
Monotony = Disconnected Teams   57
Systematization is the Mother of Progress   58
Expertise is Married to Accuracy   59
Focus is Completion's Best Friend   60
Confidence Leads You to Problem-Free Completion   61
Cleverness is the Mother of Amazing Unstoppable Teams   62

Chapter 7 The Right Way to Schedule Activities   65
Elements of a Schedule   65
New Ingredients to the Schedule Plate   66
Look Further into Critical Tasks   67
Factors Shaping Your Schedule   68
How to Address Unforeseen Conditions   69
A Brilliant Kick-Ass Schedule   71

Chapter 8 A Challenging Project Story   **73**
What Creates Unwanted Challenges in a Project   73
The Crisis Manager   75
Watching Conflict of Interest Closely   77
A Project Going South   81
Am I to Save, Fix or Pull the Plug?   82

Chapter 9 Success Stories in Project Management   **85**
Definition of a Successful Project   85
The WOW Factor Everywhere   86
Project Delivery with a Mission   88
When the Goal is Clear – an NFL Project   89
A Dream is Realized - MIT   91
PM Benefits Your Business Profitability   93

Chapter 10 How to Assess Your Results   **97**
Every Project is Uniquely Different   97
A Measuring Stick   99
Useful Life Costs   100
Close Out   102
Sustainability   103
Decode, Decrypt and Decipher All About It   105
A Final Note   107

About the Author   109

To all of you who manage anything,
This one is for you.

Foreword

This eye-opening book is written from Gustavo Valenzuela's perspective and his extensive personal experience in architecture, engineering, construction and project management. In fact, *The Book on Project Management* is a clever compilation and the sum of his project management experiences from diverse projects in which Gustavo effectively participated managing and guiding stakeholders to an indisputable successful completion.

You may be managing all types of projects in your life, I eagerly invite you to read this amazing book on *Project Management*. Especially, if you are a project owner or developer that has never truly benefited from working with an experienced project manager, this book is written for you. Whether this is your first project or your one-hundredth project you hope to effectively develop, this book is most definitely for you and for everyone in your organization.

Many chapters in this book will redefine your understanding of project management and guide you to stay focused on the authentic project's mission and meaningful goals. Even if today, you consider yourself an experienced project manager, your knowledge, skills and abilities will be ingeniously elevated to new levels. As a current manager, you will certainly enjoy learning new ways of looking and managing people, their skills and assigned tasks.

Gustavo is a true project manager at heart. His architectural education and first-hand experience owning a construction company and acting as general contractor have complimented his extensive knowledge in project management. Consequently, he is often hired to manage experienced teams composed of architects, engineers and contractors particularly, unique stakeholders with ambitious projects deemed by most difficult to develop.

Recently, we held an honest conversation about the importance of sharing and teaching others his project management skills. I honestly admire Gustavo for fully committing and taking the time to write this book to share his knowledge and unique point of view with you.

Enjoy this fun book on *Project Management* and get ready to become more efficient as you improve your organization's bottom line and increase profitability.

Raymond Aaron
New York Times Bestselling Author

Acknowledgements

I would like to thank every single one of you who inspired me to always give my best professionally, and who strategically reminded me what I am capable of accomplishing.

Many individuals of diverse ethnicities, and of different age groups, have gracefully crossed my path in my 25-year career while working in architectural firms, engineering firms and, during the past 15 years, as an owner of several businesses in design, construction and project management. The content of this book on project management is certainly the sum of all experiences gathered during this extensive and meaningful professional journey.

Special thanks to **Robert Cambridge,** from Cambridge Construction, who taught me valuable lessons regarding *planning ahead* for an efficient day of work at the construction site. Thank you for instilling in me the importance for architects to fully understand the diverse construction techniques and, more specifically, the importance for being considerate while designing any buildings in response to established processes to the American way of constructing buildings. Working every construction trade to build projects alongside you, and teaching with you at Pima Community College in Tucson, Arizona, several construction trades, including Masonry, Concrete, Framing, Plumbing and Electrical Construction Techniques, was a solid foundation for my educational and professional journey in architecture, engineering, and construction (A/E/C).

I would like to recognize faculty and staff at Pima Community College, and the University of Arizona, College of Architecture for gifting me with knowledge and tools to embark my journey in the A/E/C industry. I feel that the best U of A faculty and students of all times were present during the 5 years of my architectural education, and I would like to sincerely recognize them all, especially **Chuck Albanese**, **Richard Larry Medlin,** and the **Architecture Class of 1997,** for taking part in this highly important phase of my career and professional life. I am extremely grateful for sharing five years of intense and enjoyable architecture learning in the studio with amazing architects in the making, including **Kent Miyake, Christopher Gerber, Gannon McNeil, Chelsea Grassinger, Yasser Malaika, George Zazueta, Brian Haines, Edward Vergara II, Ryan Smith, Michael Bennett, Jason Strodl, Scott Nugent, Justin Kerfoot, Lee Petit, Jason Szeman, David Curtis, Brock Grayson, Michael F. D'Andrea, Jennifer Betancourt, Jeff Fisher, Sitara Wilson, Jeff Simutis, Brent Hoefflin, Jake Mennella, Jim Ransco, Scott Smart, Brian Gee, Diane O'Connor, Gabrielle Harlan, and Hussein Cholkamy.**

I would like to express my sincere gratitude to **Arturo Coppola** and **Juan Jaime de la Torre,** at Esquema Architecture, for employing me early during my educational journey. Attending school, and working in your firm, allowed me to experience architecture in the classroom and in the real world. Your teachings truly complemented my journey, and catapulted me to handle large, complex commercial projects while working in several firms in the USA.

Thank you, **Andrea Forman,** from Forman Architects, in Scottsdale, for trusting my abilities— fresh out from college—to handle a myriad of complex tasks most often handled by senior architects with many years of experience; working alongside you

Acknowledgements

was truly a blessing. Also, thanks to **George Pasquel III**, at Withey Morris, for participating along my journey.

I am grateful for working with so many professional individuals who truly understand the value of assisting project owners in a way that is meaningful and productive: especially **Rene Moreno**, at Piper Jaffray; **Fred H. Rosenfeld**, at Gust Rosenfeld Attorneys; **Charles Van Block**, at CVB Architects; **Ben Ferguson**, at FERGUSON Architecture; **Danielle Hunt**, at West-MEC School District; **David Schmidt**, at DLR Group; **Eric Gilliland**, at Gilliland Design; **Jorge & Bernardo Ramirez**, at JLR Designs; **Ryan Adams**, at Honeywell Aerospace; **Michael Dulberg**, at Burch & Cracchiolo; **Ted Hawkins**, at City of Casa Grande, **Chris Harlien**, from Arizona Luxury Lawns; **John Hamilton**, at Allegion; **Carlos Naranjo**, at Todd & Associates Architects; **Luis Rodriguez**, at L&R Roofing Solutions; **Matthew Miner**, at Kaco; **Scott Harter**, at Desert Recreations; **William Rathsburg** and **Jon Schaffer**, at MercatUS International; **JB Pham**, at SmithGroup; **Yash Chaudry**, from Arcore Group; **Jonathan Hille**, at Hille Companies; **Scott Wolf**, at Wholesale Floors; **Ben Barcon** & **Scott Waltenburg**, from ADM Group; **Brian Welan**, **Barb Garlow**, and **Heather & Scott Wilcox,** from Interior Solutions. I would like to express my gratitude to **Arturo Carrizoza** for reminding me about the importance of giving back to the community and making charitable donations including time and services to assist schools and organizations without financial resources to be guided through the complexities of project development.

My sincere appreciation to **David Hunt** and **Tamara Caraway,** from Hunt & Caraway Architects, for making me part of an incredible efficient and creative architectural team to design and build schools, public safety buildings, and religious centers. Special thanks to **Jonathan Schmid, Ken Robertson,**

Tammy LePham, **Leo Veldhuizen** and **Tatyana Armstead** for working side by side in a fun, effective, and productive way throughout project development.

I gladly thank **Ari and Nancy Acka**, from ANA Consulting and Structural Engineers, for working closely and diligently with me while designing residential and commercial structural systems. As a team, we designed and created remarkable working drawings with great ease, and without any RFI's, during the construction process. Working with you as a team, I learned the value of detailing diverse types of buildings in a way that contractors can clearly understand and correctly build.

I sincerely thank and appreciate **Vispi Karanjia** at Orcutt|Winslow (OW) for providing outstanding professional architectural services and for genuinely listening to the client's goals and aspirations. OW's commitment to providing excellent architectural services and addressing the true needs of the project, creates the right environment to find the right solutions. Special recognition to the amazing, fun and committed OW's team including **Jesus Fernandez**; **Liz Graves**; **Mike Nandin**; **Veronica Gonzales**; **Rahib Sarela**; **Teodor Paul** and **Chuck Kottka**

I would like to also recognize all engineers and engineering companies who have worked diligently side by side during the development of many projects: especially **Aaron Ament**, from AEC Engineering; **Doug Osborn** and **Percy Myron**, from Hess-Rountree; **Jeff Wimmer**, at Dibble Engineering; **Greg S. Broderick**, at Broderick Engineering; **John Echeverri and Cesar Reina**, from EJ Engineering and EJ2 Construction Group; **Eudjen Savu** and **Rodney L. Hillis**, from Arizona Pinnacle Engineering (AZPE); **Temo Gracia**, at Gracia Engineering; **Alex Verdugo**, at Kramer engineering; **Tom Harris**, at Sky

Acknowledgements

Construction & Engineering; **Daniel Hinojos**, at DEIH; and **Peter Takach**, at Takach Engineering.

Special thanks to the many contractors who have complemented my construction experience: particularly **Shane Alexander**, from Alexander Building Company; **Gabe Ortega**, from Sletten Construction; **Bruce Balls**, at B&F Contracting; **Brad Walker**, from DL Withers; **David Gray** and **Robert Watley**, from DNG Construction; **John Montalvo III,** and **Robert Perez**, from United Southwest Construction; **Gary Denton**, from CS Construction; **Barry Chasse**, at Chasse Building Team; **Tim Tyrell**, and **Mike & Ryan Nichols**, from RyTan. Also, **Dennis Miller, Bill Cox, Shawn Maurer, Natalie Orne, Todd Steffen, Gary Wenk, Mike Stecyk, BJ Pennington, Don Frank, Leroy Trujillo, Richard Davidson, Secundino "Cundo" Lizarraga, Slade Gibson and Shamayne Rustebakke**, from CORE Construction; **Steve Barry**, with Pueblo Mechanical; and **Mark Farrel** and **Michel Farrel Rojas**, at Progressive Roofing, **Wade Woodruff, Randy Leaphart and Paul Manos** from Woodruff Construction. Thank you all for working diligently as a team to find solutions and for always keeping the important goals in mind.

It is with great admiration and respect that I extend my thanks to **Frank Gehry** for teaching me *"how to create"* anything I can visualize by simply working together with my clients. Thank you Frank for inspiring me to always listen inside of me for the correct answers! Learning from you and following your advice on how to work with a multitude of people, and seeing the value of putting a versatile team together, allowed me to promote *team participation,* and as a result empower every team member in all my projects. Your Master Class teachings have positively impacted my own project management style. Particularly, I appreciate you sharing with me your very personal take away from the Walt Disney Concert Hall project. Lastly, thank you for your remarkable contributions to architecture.

I openly and sincerely thank **Justin Rojas** and **Ignacio Gastelum,** from Urban Energy Solution (UES), for their unique, efficient, and unbiased approach to providing outstanding electrical and energy solutions. Thank you, and the entire UES team, for working diligently, and for always keeping the most important goals in mind, and for helping so many school districts throughout Arizona. The A/E/C industry can truly benefit from companies such as UES doing exactly what you do and project owners and managers can rest assured knowing they are in great hands.

There are many organizations, and amazing individuals running them for the benefit of the A/E/C community and, therefore, our own personal growth. I would like to personally thank all of you for your contributions, dedication, and cleverness: especially **Jeff Weiner**, at LinkedIn; **Frank Stasiowski**, at PSMJ; **Mark A. Langley** and the entire organization, at PMI; **Robert Ivy**, at AIA; **Jay Elbettar**, from ICC; **Keith F. Williams**, at UL; **Jason Boyer,** AIA Phoenix Metro; **Rick Fedrizzi, David Gottfried,** and **Mike Italiano,** at USGB; and all software developers, including PROCORE, Wrike, Miscrosoft Project, Workfront, Deltek, SolidWorks, Autodesk, Mavenlink, LiquidPlanner, Targetprocess, workzone, Sketchup, Trello, JIRA, Zoho, Podio, and Easy Project—thank you for providing amazing support and giving us incredible tools to support our project development efforts.

I would like to thank RESD#2 Governing Board: especially **Joyce Luckie,** **Dr. Jaime Rivera, Ruben Gutierrez, Jose Moreno and Stacey Hawkins.** Special recognition and my sincere appreciation to **Teresa Solares, Zorina Gray, Andy**

Acknowledgements

Valdivia, Joseph Moreno, Alutha Johnson, Marcus Pina, Rochelle Elliot, Maria Carbajal, Paul Perez, Talmadge Tanks, Celia Trujillo, Anna Villa, Joanna Cordova, and **Ramona Gonzales,** for six years of amazing team work, focusing, and working in many life-changing projects. Together, we have made a positive impact on the riverside community and its deserving elementary school students. I am humbly grateful to you for choosing me and trusting my abilities as your expert project manager.

Special thanks to **Rosa Saenz,** from The Professional Group Public Consulting, Inc. (PGPC), for working closely and meticulously with me to procure for the best team, in full compliance with strict rules and regulations, and for treating every project with the same level of detail and attention, to ensure its success.

Lastly, I get to acknowledge you for holding or digitally reading this book, and for taking the time to advance your project management skills. May your project management journey be meaningful, safe, and exciting.

CHAPTER 1
WHAT IS PROJECT MANAGEMENT?

Official Definition for Project Management

First, you must consider yourself a manager—more specifically, a *project manager*. You manage many tasks, activities, and projects of all sizes in your daily life; and perhaps at work, if you either own a company or are employed, and trade your time for money. Go back in time—as far back as possible—and recall when you began to interact with adults, and learned to follow your family rules and a daily family routine. I invite you to remember your education school days and how you were taught lessons in diverse topics, and given certain tasks and exercises to complete, including homework and study requirements to prepare you for a test in order to successfully complete all that was required of you to move to the next grade. All of these activities with very specific goals allocated within a time frame were managed by others, such as a school teacher in the classroom setting example, and in other instances planned and executed by you at home, such as the case of getting your homework completed. Any of the previously mentioned examples could be considered a project that started at point A, and the goal was to advance it to Point B. You may also think of it as starting and finishing a project or, in the education example, as starting a school year and completing all assignments to move to the next grade.

It is a matter of focus, or a choice you may consciously make to look at every personal task, goal, dream, and even every challenge coming your way, and view it as a project you strategically plan and committedly work to complete. And now that you have read some basic life comparisons, the list of examples you may enlist right at this moment, so you openly accept the fact that you are indeed a project manager managing several personal projects, may be endless; it is certainly on-going. Having stated this, your very own definition of project management is rightfully shaped by your own story and your unique way to handle anything that has shaped you and has allowed you to move forward in life as you complete your very own projects. Your experiences, including all disappointments and successes, have all come together to allow you to become a well-rounded and a better, more experienced project manager.

Professionally speaking, the official definition for project management, as written by PMI (Project Management Institute), reads: *"Project management is the application of knowledge, skills, tools, and techniques to project activities to meet the project requirements."* This definition, and the basic meaning, is simple to understand; however, if you dive deeper, ask yourself: *what is considered knowledge in my industry? What are the required skills, and what tools and techniques are the correct ones in order to meet and successfully complete the project requirements?* Answers to these questions wisely come to you after managing diverse projects in your field for many years. Acting as a project manager, for many projects, has added to your extensive experience as you take them from start up to completion, while always meeting the requirements clearly defined at the beginning of the project.

The rules may be the same for every project, intending to simply and efficiently move from idea or concept to successful completion; however, the management style of any given project may be unique and rightfully influenced by the experience and expertise of the project manager or project managers. Understanding diverse personalities for the team members, and recognizing that every stakeholder may have their own version of how a specific project is to be developed and completed, is also an ability project managers develop. With time, you learn to master the managing of all tasks and people in your team to successfully meet the project requirements.

You may be asking yourself: *so, what is the most successful or the correct project management style to use in order to excel at every project?* The answer will be briefly explained in the next paragraphs. Keep an open mind as you read on, and the rest of the chapters in this book will also complement your project management understanding.

Project Management Styles

You must agree that you are a unique person who has been shaped and formed by all of your own life experiences, many of which have helped define your personality and the style or type of person you are today. This concept also applies to project management and your exposure to similar or diverse projects, ranging in scope, budget, and complexity, which will build your experience and expertise. Thus, this allows you to fully understand every aspect of what is required to successfully develop a project from start to finish and, in many cases, take it beyond into performance evaluation, warranty periods, and sustainability requirements, as required by the project's life cycle.

Managing a project requires very specific and unique skills, all of which must be part of any project manager's arsenal. Skills in negotiation, decision making, trust building, crisis and conflict management, and great verbal and oral communication allow you, as the project manager, to remain in control of the entire team and their performance. One skill, often overlooked, is the ability for any project manager to be able to see the whole picture. Being able to visualize and fully understand the entire project is the best ability you may have and utilize, in order to know immediately, at any phase and at any given time, if the project is moving forward in the best possible way. Being a strong leader, or leading your team, is the ability to simply allow them to see what you see, and to guide those in the process who are not able to see where and how the project is to be advanced. Seeing the whole picture and leading your team is therefore vital for the successful completion of any project.

Different styles of running a project are required and are often the result of the project manager's authority and assigned roles, especially if you, as the project manager, are in charge of the budget, and you are responsible for making decisions that may directly impact the project's finances.

The *Authoritative Management Style* is usually employed by project managers who are very experienced and base their decision-making process on their proven outstanding results and past successes. This particular management style is knowledge and experienced-based, and the entire team is very respectful of the PM's credentials. Every team member willingly and openly recognizes their limitations or lack of similar experience as they work to contribute via their own assigned tasks for the benefit and completion of the project. An example for this style would be that of a NASA project manager who is in responsible charge of handling a project to develop a spaceship to fly into outer space, land on the moon, and later return to earth, safely.

This NASA project manager may utilize an *Authoritative PM Style* if he or she has lived this process, and successfully completed every requirement of the mission, perhaps not only once but several times. The team members working on this NASA project, although highly qualified, may still be lacking critical credentials or specific relevant experience in order to suggest or influence the processes. Team members and stakeholders, in general, are still able to contribute and participate using their individual strengths and skills; however, they are always displaying deep respect for the decision-making process and recommendations of their authoritative project manager. The environment is typically of high respect, and communications are very formal. The team's trust is instilled in the experienced project manager as the captain leading the ship to a safe place. They are all a team of experts, presenting and allowing the expert higher power to approve and evaluate every project phase for completeness and accuracy. The authoritative management style allows for the completion of the most challenging projects being developed by a project owner, who has great vision but lacks the required knowledge and proven experience to take on the responsibilities of a *master project manager*.

In a *Musketeer*, or **All for One and One for All** management style, the project manager provides almost no assistance or supervision of the process or individual team efforts, and holds each team member accountable for deciding how to work and complete their assigned tasks to meet scope, budget, and schedule. This management style allows team members to work at their own pace and, as a result, lacks a strong leader to have a close eye on schedule; thus, lengthening the project completion time. The emotional wellbeing and overall energy of the project is very relaxed, with minimum sense of urgency, and team members do not feel they are truly being managed; they feel permitted to provide suggestions on techniques or processes to deliver the project, and they invite and promote interaction with all team members. The final results at project completion often do not exactly match those as originally stated in the project requirements. Furthermore, costs and schedule become a struggle when using this specific management style. There may be projects where this methodology is beneficial; and you, as the project manager in charge of development, can use your experience to decide when is most effective.

> *When I personally get hired to assess and work closely with teams at architects' offices, I spend some quality time documenting their existing practices, in order to evaluate their efficiency and radically complement their own processes. Then, I create a detailed plan, and I train them individually or as a team to implement useful techniques, which in a nutshell, will advance their methodologies to levels never experienced before.*

The **Pacesetter Management Style** relies on setting and holding every team member to very high standards, and often makes drastic decisions to eliminate or remove team members not able to keep up with the pace of the project. This specific style is robotic-like, and there is never free time to do anything else other than getting the work done, resulting in a high stress team environment. Team members do feel a strong sense of urgency throughout the entire process, which causes no interaction, and limits the communication amongst team members. A project manager utilizing this particular style is very knowledgeable in processes or task duration, and understands how much time is required by each task.

The project schedule is micro-managed, and delays can be identified almost on an hourly basis. The team feels disconnected from the end results, and there is no sense of pride in completing the final project. Budget and schedule are always on track, and the scope, as identified initially, exactly matches the final completed project. There is very little to no variations throughout the entire development process, as the team members tend to be experienced for the task they are required to complete. For an extreme visual example of what this process may look like, watch a brief segment of a movie where a soldier asks a worker to manufacture a specific part, while he tracks the time required to make just one piece. After the worker has completed the piece, the soldier then looks at the quantity of pieces completed and compares them to the number of hours the worker has worked in the day. The soldier then proceeds to remove the worker to execute him outside the manufacturing plant. Although this example is extreme and very graphic with the level of stress that may be created, accessing erroneous techniques and personality traits may be equally as stressful with you acting as the project manager. Very few project managers possess the personality traits required to promote trust and keep a positive attitude while utilizing this particular management style.

The *Coaching Management Style* bases its success in handholding of stakeholders and team members for the proper understanding and execution of tasks and phases completion. The entire management process is open-book format in order to provide complete understanding throughout all the established known and unknowns. You, as the project manager, if using this style, are required to become one with each team member as needed or requested throughout the entire process, including during downturns, challenges, and failures of the project. As you utilize this coaching management style, you must become an expert in understanding and managing all project risks. It is of high importance that you understand the diverse personalities of every team member in order to keep the project moving forward towards successful completion. Often, members with inadequate knowledge or skills are accepted to participate in the team so they may gain the required experience. The leading project manager is extremely knowledgeable; he acts as an instructor to ensure each team member feels safe participating in any given phase, or for some specific stakeholders to be engaged throughout the entire development process.

The scope of projects where this Coaching Management Style may be employed is usually in projects with simple scope, which rarely requires expert advice or specialized skill during any of the development phases, including the initiating, planning, production/execution, or the closing phase. The project completion schedule has abundant time to allow teaching and learning of all stakeholders, and ample time is especially allocated to the production and execution phases.

Often, when a new project is created, several team members feel that they too can act as project managers due to the exposure or open-book format. However, since every project has unforeseen conditions, and new challenges always arrive, making difficult decisions affecting scope, budget and schedule, and complying with diverse unforeseen legal requirements, it will create a sense of insecurity and instability, as *feeling* you can be project manager, and *being* one, are two completely different things.

Any of the above listed management styles, or a hybrid of two or more styles, may be the right fit for your next project. Your personal leadership skills and personality traits, while acting as the project manager, play an extremely important role in the emotional well-being of the team in order to gain their trust, which directly affects the successful completion of any project, regardless of the project management style you chose.

What is Project Management?

Your own ability to be flexible and change directions as required by the project, or adapt to unforeseen conditions while you remain faithful and true to the original defined and accepted scope requirements, has far greater value and a more direct impact than the actual specific style you select to run and manage any given project. Your capacity to create and keep trust amongst team members, and inspire them to give their best performance, is simply based on your own people skills and your own ability to understand and speak their language. Your specific way of being and managing a project is mostly based on your own past experiences, which inevitably complement and shape your personality to handle tasks and people as you promote and set the stage for a productive safe environment, in which each and every task in all project phases moves on the right path to a successful completion.

LOPPM (Lack of Proper Project Management)

Can you imagine going to the hospital and arriving in the emergency room, only to find out that the person greeting you is inadequately qualified to assess if you are in a real emergency? So you arrive in the hospital with an apparent serious emergency and eventually you are moved to an examination room where a nurse asked you some critical questions and she has made some decisions based on her expertise or experience; The nurse then informs you that a doctor must see you immediately. The doctor finally shows up, and he or she is faced with making some pretty tough decisions based on approved medical practices, hospital rules, and regulations, as well as your health insurance constrains and his medical code of ethics. The doctor then chooses to transfer you to a different hospital as he informs you that time is of the essence, and that he is not able to assist you or treat you because your condition is outside his medical practice, and it requires a specialized doctor to assess you and provide the proper diagnosed treatment or cure. All the while, you are at the mercy of their expert judgment for your suffering condition, and although you are concerned for your very own life and well-being, you hope that the doctor is making the best possible decisions on your behalf. You accept the fact, and you begin to realize that the medical profession licenses medically trained professionals to be in control of your health and, in this case, your very own life. After a line of questioning, back and forth, you now realize, even though it is your very own life at stake, you never took the time to read all of your health insurance documents to fully understand what medical care you are entitled to receive as part of the agreed medical coverage you purchased.

At this time, you also realize that doctors are somehow limited and may be making decisions aligned with strict guidelines, which at this precise moment, they simply do not benefit you. Furthermore, you sense everybody around you genuinely wants to help, but taking the required action is outside their obligation or scope to do so, and you, as the patient, most likely will pay the consequences. You become aware that your chances to remain alive and well at this point are directly impacted, affected, and limited to medical staff skills, medical insurance restrictions, and many other factors now completely outside your control.

Finally, you are transferred and seen by the proper doctor, and the results are in. It is too late and, as a result, both of your legs will be amputated. You asked yourself in a desperate loud tone, "What do I do now? How can I get out of this situation? Can anybody save my legs? Please! ...Nobody?" Then, "Can anybody please tell me who is to blame for the late medical assessment and, as a result, the potential loss of my legs?"

What is Project Management?

You, as the project manager, will face similar questions in a totally different setting, in project development. When you manage complex projects, you will certainly be faced with far more complicated questions as stated above, and especially when you are brought into an on-going project, which is lacking the watchful eye and expertise of a seasoned, experienced project manager. In many occasions, project owners can only appreciate you when the project is in crisis, and they are desperately seeking to shift responsibility to somebody else for their poor decision making and inadequate experience managing their project. The dramatic medical example is just to prove that people and project owners will embark and take on projects without fully understanding what they are getting into. They will sign the contract, ignoring the intricacies of every project and its many requirements and processes you must follow in order to deliver that project within scope, schedule, and budget.

In the medical example, many individuals ignore the limitations and their rights, outlined and fully explained in writing in their very own health medical insurance contract, which they bought but never bother to read and properly understand. And consequently, in this instance, it's their very own life at stake. Expect project owners to risk developing projects without having the benefit of a project manager who can save both of their legs, and their lives, when the project enters into crisis.

Lack of client representation via a project manager is like choosing to board a plane with a person who only claims that he can probably act as pilot to fly the plane, and it is unknown if this person knows how to assess if the plane is safe or if it has any major problems. Does he or she have the certified knowledge to investigate if the fuel and navigation gages and devices are fully operational? If you are in a position to develop a project, and are reading this paragraph, know that it is just as bad to go without a project manager. And, as explained in the airplane example of boarding a plane, which may not properly take off— and if it does, it may not properly fly— once you are up in the air, it is uncertain and very unlikely that you will reach your desired destination.

✚ *When personally working closely with project owners, I concentrate to increase their understanding during the entire process of project development. As a result, they are thrilled to feel connected of their project, especially having full control of their budget. They truly appreciate having my experience and expert advice as I act as their project manager. If you would like to hire the services of an experienced project manager, please visit* **www.thebookonpm.com**, *to schedule and receive a complimentary introductory session.*

Risking not properly defining scope, and entering into the production and execution phase to manage every aspect of a project, including moving to properly close out a project that follows a certain schedule and meets a budget, is irresponsible and unacceptable practice for a project manager. But why would it not be for decision makers who decided to develop a project, choosing to ignore proper scope definition, leading to financial disaster and a broken schedule, and wasting money, time, and resources? If you are in a position to make a decision to include a project manager while developing a project, regardless of its complexity, speak and take action to include and hire the best project manager your budget can afford.

Who Runs Project Management?

Businesses, in general, are organized with different positions and responsibilities in order to identify the level of obligation and accountability for each member or employee as they strive to run their daily tasks to stay profitable. Understanding responsibilities in any organization allows to set accountability parameters; to say the least, it paints a clear picture for the person in that specific position, and outlines precisely what is expected from them during the process of working and completing their assigned tasks.

It is crucial for you, acting as a project manager, that you fully understand the capacity and expertise of every single person, entity, and consultant working on your project. Having an organizational chart is of great help; however, it is equally important that your organizational chart is tied to a document, such as a legal executed contract, which explains the scope for each person in your team enlisting in great detail what is expected from them. This enables you to understand, basically, what is in their to-do list. The question of who runs project management is asked to allow you to think that there are stakeholders or entities in different positions that may and most likely will affect the development process; thus, making you feel that they are running the project in some instances. You, as the hired project manager in control of your project, are responsible for running all such entities, and for orchestrating their requests and tasks, keeping in mind that stakeholders may influence your process. Accounting for the owner or project sponsor requesting and making changes is also another opportunity for you, as the project manager, to remain in control, managing the project as you make them aware that budget and schedule will be affected by their requests. In some cases, they may want to reduce time and require a faster completion time as they shorten the project's completion date; this will affect budget and, if the budget is to be left untouched then, the only option is to reduce the project scope. In sum, the project manager is always running project management despite the many attempts from stakeholders to deviate or steer the project on a different direction.

What Is and What IS NOT Project Management

During the procurement and selection process for the best possible team, the main goal is to select the most effective and ideal team that will preferably run itself, without micro managing every aspect and assigned task. If you are successful, and your budget can afford to ensemble a team of experts, the project will most likely run smoothly, but only if you are able to properly manage it. Having the best team does not mean for you to hold a title of *project manager,* and choose to allow for things to just unfold as you remain absent from watching every aspect and process, scope compliance, budget, and of course, schedule. Allowing things to happen, and disconnecting your management eye from a project because you have established trust and confidence in the team's abilities, is not really considered project management.

Even when you successfully acquired and have the best team to develop your project, acting as a project manager, and looking for opportunities to further improve any processes, will eventually save time and money; this is definitely considered project management. Having a team of experts promotes efficiency and sets the stage to perhaps discover new ways to improve the project's bottom line. There are opportunities with every project to adjust, improve, and even systemize processes to become more efficient; and you, as the project manager, are in the ideal position to stay engaged throughout the entire process to capitalize on those opportunities. By performing your duties as a project manager at all times, your expertise and level of confidence will be elevated and rightfully separate you from amateur project managers seeking to just complete a project just as required or even for less if they can get away with it. This is not to be considered an ideal practice in project management.

What is Project Management?

In your very own response when you ask yourself what project management is, you may now add, *"the ability to remain connected and engaged throughout the entire process, in order to positively affect scope, budget, and schedule."*

A Robust Definition for Project Management

As you work in project management, you will realize the importance to take your duty as a project manager beyond your contractual agreement. You are already expected to apply your knowledge, your skills, and your techniques to meet the project requirements, and that is about the extent of what is expected from you contractually while working in the industry. Even more important, using your unique set of tools, such as the ability to recognize not only your own limitations but those of the entire team, and choosing to mitigate them by relying on expert advice, goes hand in hand with the ability to decipher who is an expert in that arena. By choice, you must make it a personal practice to consciously look for the greatest benefit of the project, regardless of what is in your project manager contract.

Project management is also inclusive of moral values, both personal and those of the stakeholders. Project managers must also have proper understanding while managing diverse personalities and cultural differences and beliefs. Your decision about going beyond just management of project scope, budget, schedule, risk, quality, and resources sets you apart, as this practice of going the extra mile is only for a select few; nonetheless, it is an important part of project management.

How will you know when you are going the extra mile? Simply when you come to that situation in which you tell yourself, "This is outside my contractual duties," but you are certain that if properly addressed and handled, it will positively influence the outcome of the project, and then you simply choose to do it.

 *To download guidelines on effective steps and ideas on how you may push yourself to consistantly go the extra, please visit **www.thebookonpm.com**.*

Lastly, in a new definition for project management, you could easily add: *"the ability to include everything tied to any project, and place it under the responsibility of a qualified Project Manager (PM), who is able to handle everything, always taking action for the benefit of the project."* That would be an all-inclusive contract that only a few project managers would confidently sign. If you include yourself in that list of the selected few who would sign such a contract, you are a project manager at heart, performing project management in its entirety.

Chapter 2
Why Project Management

Projects Developed by an Inexperienced Workforce

Who can develop projects? You guessed it: those who want to and have the financial means to do so. Now, can anybody play the role of project manager, and run a project? They sure can; however, the results will not be favorable for any of the variables that matter, such as scope, budget, and schedule—especially budget. They will lose money in the millions as a result of their innocence and obvious unsophistication in the subject. You may be surprised how many projects in the world are developed by an inexperienced workforce and, in most cases, without the benefit of having an experienced project manager. In the United States, around 68% of the projects fail in at least one of the three critical areas, and 98% of those failed projects lacked the services of a project manager. Even more revealing, the remaining 32% of the projects that did not fail were mainly because they were properly organized and managed by an experienced workforce, and they included the services of a seasoned, experienced project manager.

Now, come back from the worldwide scenario, and zoom into the USA and, more specifically, look at the education sector. Pick any of the 50 states, and choose public schools. Let's say you selected Arizona: it is very common to find public school districts with enough capital, in the millions of dollars, to remodel or build new schools. Many school districts in Arizona have the need, the desire, and the funds to remodel and build state-of-the-art schools. This is great news for students; however, keep in mind that the entire public school district is composed of educators and business people focusing on education. As a result, they lack the required experience to engage and properly manage architecture and engineering firms, general contractors, and a handful of specialty consultants, who are so specialized that nobody in a public or private school organization would even know these contractors actually exist.

Still, many school districts choose to develop projects without having a seasoned, experienced project manager who knows all the ins and outs of every stakeholder and potential team member.

Even more disturbing, many school districts *do* understand that having a project manager to represent them, and handle all that is required by project development, entirely eliminates the risk of making costly decisions; nonetheless, they still choose to *not* have one. But how costly is it, you may ask?

For every billion invested in the United States, $122 million is wasted due to lack of project performance and proper management. Even more shocking, the failure rate is 50% higher for projects with budgets over $1 million. In most cases, the project owner does not even know they are losing money, or that they are *overspending*.

Next time you are acting as a project sponsor, or if you are the actual project owner, make the right choice, and avoid pretending that you have the correct talent in your organization. Additionally, avoid assigning your project and its proper

Why Project Management

management to an inexperienced workforce.

Effects of Having an Experienced Project Manager

True fact: on a 17 million dollar project, a project manager, managing the development of a new state of the art STEM school, saved the project owner over $3 million dollars by reviewing each separate line item during project estimating, and by working closely with the A/E/C team to demonstrate how construction costs on the final GMP Design/Build document could be adjusted without modifying the original project scope. To accomplish such significant savings, ranging in the millions, an honest open-book look at all costs was performed.

In the same project, over $1.3 million dollars in additional savings were achieved by procuring and separating furniture into three separate categories, and allowing the direct vendor for each category to provide direct pricing to the project owner. In sum, over $4.2 million dollars were identified as savings in a project that lasted only 14 months. The salary for the project manager was capped at $300k, leaving a total savings in cost reduction and adjustments of at least $3.9 million, in the project owner's favor. Scope and schedule were met as approved by the project owner and sponsor. Could this have been accomplished without a project manager? Historical data and actual facts on projects, without an experienced project manager, leads to the answer of a resounding, "No Way!" It is of great important to note that an impartial project manager is one who is only representing the interests of the project owner and must not be a project manager wearing multiple hats or working for other trades. To be more specific, the project manager is to hold a contract directly with the owner and not be an employee of any of the other trades. Often, architect's or general contractors' developing buildings have project managers within their organization and may offer to the project owner to act as the project manager for the entire project. This scenario is to be avoided at all costs, as it is allowing a single entity to basically write their own checks and cover up their own mistakes.

There are many benefits of having an independent and unbiased project manager. Proper and correct disbursement of financial resources is one of the positive effects of having an experienced project manager managing every aspect of any project throughout the entire development process. Full compliance with contractual obligations and approved project charter is another major plus when you, as a project owner, get to benefit from having an experienced independent PM as the watchful eye of your project. Assessing changes in scope, schedule, and budget, and how they may affect the goals of the project as identified by the project owner, as well as mitigating potential challenges, thus avoiding a crisis that may hurt the outcome of the project, is more than a benefit—it's peace of mind for all stakeholders.

Having a seasoned project manager on your development team, managing every aspect of your project, is the equivalent of having a cardiovascular surgeon perform open heart surgery on you rather than the chief executive officer (CEO) of the hospital. If the comparison does not make sense to you, most likely you are a project owner, willing to live and deal with the negative effects of not having an experienced project manager on your team of experts; hence, money simply goes down the drain. The positive effects of having in your organization an experienced project manager are many, and one of the most important ones are never mentioned or credited, and that is having peace of mind that the best and appropriate person is handling the task of managing the project and will always manage and make decisions to benefit the entire organization.

Losing Money

As a project owner or sponsor, from the moment you begin to engage people and resources without the assistance of a project manager, you already begin losing money. Money is lost in project development on paper, before you actually have the opportunity to process any invoices and officially pay your project bills, or process any formal payment applications. It is common to see project owners approve contracts for consultants, services, or products without fully understanding what they are really getting into. Most, if not all, project owners, ignore the value and cost of services and products required to develop any project; even when they obtain multiple bids for such services and products, they still lack the ability to interpret costs and quoted services. Often, money is lost on unnecessary services, things, or tasks contractors omitted in their contracts, even though they knew those services were required for the proper development of the project.

Losing money on a project is always tied to ignoring key pieces of information, excluded or overpriced by the contractor, and which are crucial for a successful project completion. The indicated impartial third party is you, acting as a project manager, and one of your duties is contract writing or contract review. You must also have a well-developed ability to review and negotiate bids, quotes, or any costs associated with the project. Other key tasks, which must be performed by an impartial party, such as a PM representing the project owner, is to review the entire project scope and approve it for the correct and just amount of money. Project owners ignore how to manage scope, budget, and schedule. As a result, losing money may also happen when scope is changed without the owner's approval, or by extending the schedule and perhaps preventing the start of operations on the agreed date, which basically translates to not being able to operate your business on the agreed completion date. This translates to loss of revenue and unnecessary payments for operating costs, even though your project may not be ready for occupancy.

You, as the project manager, have the privilege and duty to know all about the financial situation of the project; therefore, it is your responsibility to work closely with the financial and business team to approve payments per agreed contract while you verify scope compliance, and approve the quality of the work as delivered. Any discrepancies in invoices vs. approved amounts and balances are to be watched closely throughout the entire project.

 To download guidelines and a checklist on effective steps you may take to avoid or stop losing money, please visit www.thebookonpm.com.

It is a great challenge to have any project scope or schedule change without affecting the budget and, eventually, the project finances. It is even a greater challenge to recover any lost funds without the presence of a project manager as the main responsible go-to person, or as a key part of your development team. When funds begin to be utilized for unforeseen conditions or unexpected activities caused by poor management, or perhaps because contractual obligations are ignored or, even worse, not known, your project starts losing control.

Losing Control

The margin for keeping or losing control of your project is very narrow. The most common factor triggering the loss of control in a project is lack of daily and proper communication amongst stakeholders; especially between you, as the project manager, and the production team engaged during the active or execution phases. You, as a project manager, must have critical project tasks memorized, and your daily checks on important aspects of the project must be performed and properly attended. Client meetings to provide project updates and periodic formal and informal meetings with your team are great indicators that you are indeed in control of your project. There are a number of factors that are directly related to losing control of your project: scope creep, departing staff, and even wrong skill sets allocated to the project, or even lack of focus from the project team, can all lead to disaster, and to you losing control of the project. This may end up causing a dent in your budget, or delaying the schedule, which in most cases is synonymous with losing money.

Managing tasks and keeping an open communication line with your active team members, and setting the stage for the upcoming team members, are all good practices for staying in control. Many factors outside your control may also cause you to lose control of the project; if that is the case, your immediate attention is required. First, take a step back and reassess the situation as a whole, and how it affects the whole picture. Once you have gained complete understanding and a cleared picture, you may develop a plan to begin turning the project around and getting it back on track.

When I become an exention of a project team to share my expert project management advice, I focus on supporting them by identifying effective techniques to avoid losing control of their projects. Then, I share effective tools and techniques so they may remain in complete control of their projects, all the way through successful completion.

Second, communicate your plan with the rest of the team, and establish a process to implement it. You may find that resourcing is necessary, and moving your best team members to address the current situation may be required in order to make the greatest impact on fixing the problem in the least amount of time.

Third, take a closer look at task management, and assess if breaking down tasks into more detailed sub-tasks is beneficial to your management reviews for compliance and completion. By checking tasks, you are also verifying established processes, and assessing if the right procedures and established protocols are being followed, including how tasks begin and are completed to comply with original scope.

Lastly, use proper measures while assessing any project challenge, and when providing a solution, be very detailed in your assessment: using facts, actual days or even hours lost, how to recover for lost time, and what tasks are essential to implement the solution. Evaluate cost and schedule, and how they are affected or corrected due to losing temporary control of your project. If no life is in danger, and safety is not at stake, always present a problem to a client when you have already identified several workable solutions, and you have all the information gathered and documented, ready for them to review and approve. The moment you start thinking and working towards finding a solution, you have already began gaining control of your project.

Pretty Schedules = Unwanted Results

As a project manager in responsible charge of projecting activities, you create and review schedules for the benefit of listing and organizing what activities must unfold for a specified duration. Then, you organize tasks in the correct order they must happen to advance the project towards successful completion. Your schedules, and the schedules you review from other team members, must be as real and as honest as possible, and they must be based on known facts. Effective schedules are never based on hope or a wish list of how things are desired to unfold. Spending your time making a schedule pretty or attractive deviates from its main objective, which is simply intended as a tool to aid and guide the entire project.

Schedules are *alive and breathing* documents; consequently, they are required to be updated and adjusted periodically. A pretty schedule—in full color and with perfect margins and format, simply created as a result of an owner's request or a project requirement—if used to run a project, will only yield unwanted results. Those schedules that you only delivered once and were never updated are most likely just pretty schedules, and they are not to be considered a useful reliable tool for anybody in your team to utilize. Although pretty schedules may serve a purpose, such as being part of the loan package submitted to a lending institution to secure the project's financing, they are not to be considered as a tool during the production and execution phases.

Relying on your memory to gage the project's progress or verify actual status of a project is not recommended; in fact, the complete opposite is the main reason why schedules were invented. Today, with the use of technology and most software applications, which can be conveniently accessed from the cloud or a virtual server, it's possible to have schedule information with you at all times, without the need to carry a printed version of your schedule. The project's result you strive to get can be mapped in an effective schedule; if tasks and activities are real, and based on actual facts, your schedule then becomes your best friend, and a great tool. It is always the information found in your schedule, and never the way it is presented, that really matters once you are immersed in production.

Later on in this same chapter, and in greater detail in Chapter 7, additional information will be outlined, which also applies to important elements of a true, effective schedule, and how, if properly used, the entire team will gain your trust, as you will be able to hold effective meetings and share detailed critical information regarding their portion of the work. Most of this information was never considered in past projects. You may even, and will most likely, become their hero.

Accountability of Stakeholders

You are making great progress, you feel in control of the project, the entire team is on fire, and you begin dreaming of finishing this complex project ahead of time. Coffee that morning tastes even better as you visualize the project being completed early, and surpassing every stakeholder's expectations. Your experience and expertise are paying off, and it is inevitable that you feel in complete control of the project. The team's moral is high, and even the governing agencies having jurisdiction over your project are collaborating and are in line with progress. Then you hear a ding: you proceed to check your email in your phone, and you read a request from the owner wanting to change and add to the original scope. Your heart is pounding so fast; it feels like it is going to come off your chest. You take three deep breaths and count to 10. Then, you read the email again, and count to 10 again. You are the only person who fully understands what the owner's requested change entails, and how it will translate to budget change, schedule change, and events coming out of perfect synchronization. You realize that a *Change in Scope Meeting* with your team may actually lead to some team members losing confidence in your project management skills.

As a project manager, you are required and expected to execute processes, and move through project phases in a clear way so that the planning and approving process for scope is so definitive and formal that no change takes place. Hence, when any stakeholder, including the project owner or sponsor, requests a change in scope, budget, or schedule, you, as an authoritative project manager, can hold them accountable for what they have requested. People, in general, change their minds constantly; if you are one of those project managers who make challenging and impossible projects seem easy because of your expertise and highly organizational skills and outstanding people skills, owners will feel invited to propose changes anytime they feel like it.

Furthermore, owners will expect you to handle their request like it is no big deal, and foresee that you will work your magic to accommodate for their late request. But then, you remember the word, *accountability,* which you intend and must use in situations where any change in scope will deviate your project from a successful completion. And, by successful, it is in every sense of the meaning, and not necessarily by completing the project on a specific date. There is a lot at stake, and it is your duty to communicate and paint a clear picture for all stakeholders.

Now, slowly breathe for 10 more seconds, and formally and officially remind the project owner about the process for approval of scope, and pull out and present their approval document with their signed name, signature, and date. Point out the paragraph addressing *Changes in Scope,* in which, in sum, it communicates that after approval, and especially during the middle of production, any changes in scope will drastically and most definitely affect budget and schedule. If the owner is still adamant about his request, prepare the correct documents, and amend the project's budget to properly and generously compensate your team for accommodating such as late request.

Give your schedule ample time to house such requests, and inform the owner of the consequences of his requested change. Many times, the only effective way to prevent project owners from changing their minds and requesting a change in scope is to exercise your right, and hold them *accountable* for their decisions, in a calm, professional, and official manner.

*To schedule an introductory meeting, and evaluate the benefits of having the expert advice of a project manager in your organization, please visit **www.thebookonpm.com.***

Chapter 3
Phases in Project Management

How Many Phases Are found in Project Management?

As a project manager, you will go through the basic phases in any of the projects you are managing. These phases may include initiating, planning, execution/production, monitoring/controlling, and closing or close out phase. These are the general, overall main phases—or in simple terms, the whole picture. But are there any more phases as you zoom in and look closer into each specific phase? In other words, are there sub phases within each major phase?

The answer is *yes*, and if you choose to see it that way, your organizational and management skills will reach new heights. A phase is simply a grouped set of goals or requirements containing a number of steps and, once completed in the allocated time, that phase has been attained, and the project can progress to the next phase. So, it is for your own benefit, as you become an extraordinary project manager, to see sub phases within each major phase, and identify what those sub phases are in your project. You must agree that since a phase has goals in need of completion, each goal has defined requirements, and they are to be executed and completed. Now, take it a step further and create a list of tasks in order to understand that sub phase.

Your list may include the following tasks:

1. Identify specific goal requirements (also verify predecessor work).
2. Define scope for each goal (include requirements to start, produce, and deliver).
3. Identify the specific team members working in that phase.
4. Outline the steps required to complete that goal by the specific team members.
5. Identify the time required for this goal to be completed, and notify successor.
6. Create a list of deliverables in order for that goal to be labeled as completed.
7. Make a list of important documents, and collect the data that may fully describe the decision-making for the development of this specific goal.
8. Do it over again for the next goal, until the phase is completed.

You have now broken down a phase into sub phases, or a major set of goals into individual goals. In a way, you have created a daily, weekly, or monthly to-do list, to execute your job as an effective project manager (PM.) Moving from goal to goal, or from point A to point B, has new meaning. This simple technique will allow you to visualize and understand the whole picture of any project in development and, at the same time, it will allow you to see and understand the smallest detail of the project.

Learn this simple process, and you will become an efficient project manager with extensive knowledge in every phase and, in the eyes of many project owners, an expert in your area.

Going from A to B

You have now learned one simple technique to move from goal to goal, or from point "A" to point "B", and to navigate from start to finish within each goal. Keep in mind that anything in your project can be separated into having a starting point and end point. The only thing that may prevent you from starting is that your predecessor has not delivered what is needed in order for you to start. Therefore, one important task is to know who and what is before any specific goal ready to begin execution. It is a smart practice to stay proactive when managing your team, and require your team to create a list of items titled, *Requirements to Start*. So, ask the specific team members to provide you with their list of requirements so they may start their portion of the work without any excuses or limitations.

Now, let's assume your vendor gives you their requirements to start, and you go to their predecessor to ensure they are delivering exactly what is required by their successor, only to find out that what is being requested is not on their contract. This happens more often than you would want; therefore, it is vital for you to know goals in phases, and their requirements; it is truly an important piece of information to have when writing and executing contracts. Work closely with your legal and expert contract team to include at least general language in your contracts to avoid omitting vital task requirements, which may unnecessarily affect your contingency funds.

Finishing your tasks for any specific goal takes your project to point "B". As you complete this task, and arrive to point "B", have a checklist of items completed, and promote contact between team members working on all tasks and phases, especially those team members whose phases are interconnected. This practice may feel like you are micro-managing the project to any experienced vendor; however, it will feel like you are holding hands to the inexperienced team member providing that specific task. Use your judgment when to use this technique based on the teams expertise and experience. If you decide to use it, it will only benefit the accuracy and completeness of the project.

The Crucial Tasks Between A and B

Making progress and moving things along is definitely a great feeling. It is a preview or a taste of success prior to completing the entire project and feeling blissful. For every major phase and sub-phase, you are responsible for watching budget, scope, and schedule. In doing so, there are some crucial tasks you ought to perform in order to ensure the successful completion of that phase.

You may look at project management as a fast, efficient, and great looking car. Looking closely at all the parts making that car, and understanding that each part is essential to the entire car, you begin to understand that the smallest and sum of all parts is what makes it such an amazing vehicle. If one simple fuse—let's say the ignition fuse—which costs only about $0.99 cents, is burnt, then the car will not turn on and move forward. That small, tiny piece of plastic, with a conductor in the car's fuse box, has the capability to prevent the vehicle from starting and moving even one inch. In the same manner, the smallest, insignificant task, not properly completed, can prevent your project phases from being fully completed, and can drastically slow your project. In some cases, this small task may trigger your project to come to an absolute undesired halt.

All pieces of the puzzles are fundamental for the proper overall project completion and, by identifying some crucial tasks as you move between "A" and "B", you will set the foundation to complete each task within any specific sub phase. So, ideally, your crucial project management tasks are to include daily visits to document progress; and, if different than anticipated, make contact with the appropriate vendor immediately. When possible, photographs, daily reports, or emails can summarize your visit's results. Develop an action plan for all items potentially preventing the project from moving forward. It is extremely important that you make contact via phone or face to face meeting with the team members handling that specify task. Have plans C, D, and E, if you must, to solve the issue at hand. During the meeting with the vendor, inform them about the consequences of not performing their duties as stipulated in their contractual obligations. Receive verbal and written commitment from the team, review their action plan, and assign a drop dead due date to it. If the vendor or team working on that specific task is delaying your project, investigate their sub vendors or suppliers, and begin a formal dialogue to fully understand the issue. Often, you may find out that it is your vendor's inadequate processes and business tactics, and not the suppliers, causing the problems. If it is the suppliers, begin formulating an alternate option.

The idea is simple: just micro manage every detail of any task as required, and work with your team to find and implement the most appropriate solution. Crucial tasks in between A and B include your participation and close eye on everything that is unfolding. It is smart for you to be immersed in the process and micro-manage the project to prevent any undesirable turn of events. In fact, it is one of your main duties since the very first day you are engaged in the project management process.

When to Engage a Project Manager

As a project owner or sponsor, as soon as you get an idea to develop a project, call a project manager, and hire him to participate in the entire process. Ideally, if you are the appointed project manager, you were engaged as early as possible to have the greatest positive influence in the way that project moves through development.

A project manager can be engaged at any time, and during any of the phases of the project, including once a project has been turned over to the owner and is under the warranty period or in its 5th year of occupancy. Some project owners often require a knowledgeable professional to provide a reliable assessment on what to do with certain aspects or challenges of a project in their 10th anniversary of being completed.

It becomes extremely critical for a project owner or sponsor to engage the services of a project manager when the project owner requires an impartial party to represent them while managing a large team of experts on that project. As a project owner, the best decision you will ever make is to have an experienced project manager interpreting and efficiently guiding the rest of the team. Doing so creates a neutral setting, strictly focused on avoiding conflict of interest, and directing everybody's energy and talent in providing professional services without creating informal partnerships for the their individual benefit.

For example, in a design-build delivery method for a building, the general contractor holds the contract with the owner, and a separate contract with the architect. The consulting engineers work under the architect; consequently, the project owner ignores how the general contractor, the architect, and the engineers may occasionally choose to protect each other, when needed, to avoid contractual responsibilities. In other words, they may all solve their own mistakes made during their implementation process, and agree on how they can still benefit from a contingency or any other fund. This is not to say that they are corrupt and are preforming illegal activities. They may very well be entitled to legally collect funds from a line item on a budget assigned for omissions or forgotten conditions—which, in most budgets, should not be there. In any case, a seasoned project manager can keep things clear, honest, and legal, while holding everybody accountable and in strict compliance with their contractual obligations, on behalf of the project owner.

 *To hire and benefit from having the expert advice of a project manager in your organization, please visit **www.thebookonpm.com.***

PM Duties During Project Development

It is very common for a project manager (PM) to become the go-to person for absolutely everything while developing a project. At any time, during any phase and on any given day, you, as the project manager, may be contacted to attend the needs of the project. You are the official person who answers when anybody calls 911 on the project, or when he or she simply wants to communicate or learn anything about the project you are managing.

Your duties as a project manager will in fact vary, based on the actual phase the project is currently working on. Besides your basic duties of planning, organizing, leading, and controlling, there are other duties and responsibilities, which are vital for the success of the project. Expanding on your knowledge will come from continuing education but mainly via the experiences you gain managing a diverse portfolio of projects. Actual practice in the field also develops your skills; learning to only use what works best, based on specific project indicators, complements your project management techniques.

Improving your abilities and skills to manage scope, time, quality, costs, and risks logically, habitually happens after you understand projects as a whole, and after having the benefit of actually managing several projects from start to end.

Since you are the gatekeeper of the project, as well as the daily manager and the security guard, your duties as a project manager may be amended as required by arising needs of the project. In most instances, you will be the one making those amendments on your own.

As you become more experienced and are able to sharpen your skills and add more tools to your toolbox, your efficiency will allow you to manage projects automatically, without feeling overloaded. One last important duty is to recognize when things are getting out of your control, and to seek the assistance of your immediate supervisor or your support team immediately. Remembering that teamwork is essential to accomplish success, it also applies when you are in need of support.

Efficiency Definition Through PM's Lens

Efficiency is an art you must learn to master as you participate in project management. Organizational skills will come in handy, as well as completion accuracy, when you begin to encourage your team to become more efficient. Anytime resources are productively used, money is spent as planned, and the scope of the project is in line with set goals, and are being met, you can easily say that you are running an efficient project from top to bottom.

Knowing all the project details and team requirements allows for efficiency to be born. Then you, as the PM acting as a project leader and working closely with your team, can define critical project milestones. Keeping the communication line open, honest, and clear between team members and stakeholders, sets an invitation to all to address anything that, if held, can limit the efficiency of the entire project development. Having open communication allows you, as the project manager of record, to attain pertinent documentation required to move the project forward, especially any and all documents requiring the stakeholder's review and signature. Manage risk on a daily basis, and the need to address and take corrective action will be drastically minimized.

Support stakeholders to avoid scope creep, and guide them to prevent them from adding new elements to a project once it has been reviewed and approved. Scope creep damages your efficiency, and it can quickly turn into delays and chaos. As you complete a significant phase, test and approve all deliverables.

Lastly, evaluate the project as often as possible. You may strategically perform the same close out procedures on a weekly basis, or as often as goals and phases are completed. This eliminates fatal mistakes or omissions on important duties, which may, in the near future, affect the efficiency you are working diligently to attain.

 *To download a checklist so you may become more efficient, please visit **www.thebookonpm.com**.*

Chapter 4
How to Get the "A-Team"

What Matters Most During the Selection Process

The success of your project is based on the quality of your team members. There are many variables to consider during procurement and throughout your search and selection for the perfect vendor or *expert* who will become part of your team or organization. Often, a project owner may have many strict set rules and regulations; this is to enable and set in motion filtering to allow only those who may be the right fit to join the team and participate during the development of the project. The legal requirements must be met, and the qualifications for the team members and the company must be evaluated and compared meticulously.

Although every project experiences some kind of political tactics, politics must never play such an important role during the different phases of the project, especially during selection of critical key team members for your project. If some of the stakeholders are politically connected, it may help your project move along at a faster pace when seeking certain required approvals; however, it rarely makes a significant impact on your project and, most often than not, politics may even slow down your processes. For this reason, when selecting team members, being politically connected is to be at the bottom of your wish list.

There are critical factors to be considered when procuring, interviewing, and selecting the best possible team members. First, seek for relevant, proven, and verifiable experience, as close as possible or equal to the requirements of your project type. Second, look for people with a strong ability to solve problems, and verify they have the right knowledge and access to tools to tackle the most complex challenges the project may face. Problem solving abilities are a great asset and quality to rely on, especially during a project with processes that are fairly new, and unforeseen conditions are anticipated. Third, evaluate their longevity and stability in the specific tasks they are working on, and handling on behalf of the team. Lastly, ask them to list their strengths, and have them explain why they are interested in becoming part of your team. Listen for key pertinent words that align with all the items as listed above. Always check the validity of all information as submitted, and do verify their references.

Procurement Beyond Legal Requirements

Meeting the legal requirements to become part of the team developing a unique project is important and required. When searching for the perfect vendor, the legal requirements can be requested in a checklist format, and the vendor's responses can be tabulated side by side so you may quickly verify compliance with submitted statement of qualifications.

Once you have verified their full compliance with the project's legal requirements, you, as a project manager, may assist the procurement team with verifying the technical requirements and the project's special requirements. As you search for special skills or knowledge in certain critical areas of the project, enlist all those important requirements in the procurement documents, and make them part of the official request for qualifications. This will allow you to filter vendors and send a clear message of who you are really looking for to join the development team.

Choosing a vendor solely based on cost, in some public projects, is not even permitted, unless it is specified that you are seeking and will decide on the lowest bid. Hiring based on cost is very risky and not recommended. On the other hand, hiring the most expensive vendor does not mean he is the most qualified. As you review vendor information, it is commended that you have a diverse selection committee with specific expertise in different areas of the development, including scope, execution, finance, policy, technical, maintenance, warranty, life cycle, or other areas identified as important to be considered by the specifics of the project.

Developing a scoring system is also beneficial when funneling vendors to allow for the most qualified vendor, with the best score, to be distinctly exposed. After you have performed all these duties, you ought to consider receiving from all vendors a formal *statement,* or *summary of qualifications,* as part of the documenting process. Once you have gathered, reviewed, and scored all information, an interview with the top 5 vendors is highly recommended. In the next few pages, you will learn what questions to ask during a formal interview.

Statement of Qualifications

What makes a company or a vendor unique and the most qualified for your project is best described in the statement of qualifications, or SOQ. Lengthy SOQs do not necessarily mean they are the most qualified or the most experienced vendor for your project. As a project manager leading the project, you are in control of what information is requested as part of the SOQ; once all the legal and official information is listed, have a section or two in order to require information that will allow the decision committee to identify what vendor is the most qualified and the best fit for the project.

Items to consider as part of the SOQ may include:

- List of proposed team members and resumes (request to know years with company)
- Organizational chart for proposed team and their duties listed in great detail
- Previous exact or similar project experience and references
- Specific certifications that may directly support your project
- Personality test (from one of your pre-approved vendors)
- Background check or Approved Clearance Card
- Community service history
- Understanding of the project (a statement in their own words)
- Personal definition for success

It is important to list the requirements on the SOQ form that you and the procurement team feel are necessary in order to make the right decision as you select the most qualified vendor for the project. With every project you manage, amend your SOQ list as needed. When you are reviewing SOQ information, figure out an effective scoring point system for the diverse pieces of information you have strategically requested. Regarding anything you identify during the SOQ process as inconclusive or difficult to decipher, create a list of those specific unresolved items, and turn them into formal questions. Then, make a note to ask them during a formal face-to-face interview.

The Right Interview Questions

As a responsible project manager, working closely with the procurement committee, you have the duty to read all about the different vendors seeking to become part of the project's A-Team. Written information is great to have as proof of their statements; however, there may be a few things that still remain unanswered, or a few things you can only understand during a formal interview. Many project owners perform interviews, only to require vendors to verbally explain the same information as submitted on the SOQ; this is an inefficient use of time.

Efficient project management is directly related to efficient use of time and opportunities. Having stated this, utilize the interview to only ask questions that will give you a new insight on the vendor's knowledge, expertise, or specific skills. Also, use the interview time to ask the tough questions vendors are not usually asked. If you do ask meaningful tough questions, the vendor interview can become such an informative and crucial key in the process of selecting the best and most qualified vendor—but only if you are able to concentrate your efforts in learning things that are pertinent to the specifics of your project.

During an interview, questions regarding processes, and the experiencing of first hand decision-making abilities, are great questions to ask. During the SOQ, the vendor has all the time in the world to formulate an answer; however, during an interview, the vendor only has seconds or minutes to formulate and give his response. As a result, his response is real, and it informs you how this particular vendor will handle things during the development process. It is easy for a vendor to state in the SOQ that they have years of experience and can handle your project; however, during the interview, if you formulate real questions, based on actual project requirements, you will be able to tell if that vendor is truly knowledgeable and has the right procedures and tools to move the project forward.

Ask questions in all critical areas of the process: initiation, planning, execution, monitoring and controlling, close out, warranty, operational phase, and sustainability. For instance, if budget is known to have its greatest risk affected during the planning process, and you are interviewing the design team, you may ask them, without formal preview to your line of questioning, to describe their process for selecting materials during design, and to elaborate on their process to know about complete true costs. You may also ask them even more specific questions, including giving you a recent example when they used the same product or concept. Ask for key names, such as manufacturing company, installation contractor, and a list of projects that have used the same material. Carry it further and ask for a list of projects under warranty and data on the actual performance and reliability of that specific product. This line of questioning truly filters vendors, and those with real honest experience are able and willing to share their answers in great detail. Then, verify their response by calling their references.

Keep in mind that you, as the project manager, have full control of what questions make sense to ask in order to arrive at the right decision, while selecting the most qualified vendor for your project.

 *To rely on the expert advice of a project manager to guide your team during vendor's interviews, please visit **www.thebookonpm.com**.*

A Conclusive Selection

You have worked diligently as a team with stakeholders, and especially with the procurement and selection committee, and a decision has been reached. You, as the project manager, have identified all the different areas of the project: specifically, those areas that will require the assistance of professional and expert advice, to take the project through the process of planning and production, all the way to successful completion.

Your extensive experience as a project manager allows you to see the entire project, and understand the different phases and requirements in order to execute the perfect plan. The major key players to join the project's team have been identified, and a conclusive selection has been reached. Now what?

As you work with a team of experts to establish scope, you may refine objectives and define the course of action required to address the project owner's vision and goals for the project. Specialty areas may be further analyzed, and you may become aware of areas in need of expert advice. Occasionally, you may decide to directly hire a specialty vendor or consultant to aid the review and evaluation of newly discovered specialty items or tasks. The main reason for contracting this vendor directly is mainly to save money on the customary mark ups of other vendor housing specialty services, under their own and direct contract with the owner.

A good example in building development is civil engineering services, which can have a high fee, and if you add the 15% to 30% markup, the resulting higher fee is money inappropriately and unnecessarily spent. A civil engineer will provide the same services in response to the needs of the project, whether he is working under an architect or a general contractor, or directly for the owner under the supervision of a project manager. Acting as the civil engineer for the project, he is still required to communicate and coordinate his efforts as part of the whole team and organization. The only difference is who holds his contract and how he gets paid.

There are many specialty contractors and, as buildings get more complex, portions of a project that used to be handled by a single entity, are now being attended by 3 or more separate entities. Electrical engineers used to handle all the systems in a building that required any type of power, including all communications and alerting systems. Today, it is common to have an electrical engineer handling power and lighting requirements in a building, and having a specialized solar power production engineer, fire alarm engineer, burglar alarm systems engineer, audio and video systems engineer, network engineer, energy management systems engineer, and even an electrical systems monitoring consultant to collect data on how electricity is used in a building. Furthermore, electrical engineers sometimes rely on expert advice for lighting controls and specialty illumination requirements. Most of the previously mentioned entities, on many occasions, offer their services as a design-build option. This means that their costs include the design and the installation of that particular portion of the project.

Design-build delivery option is one of many different project delivery options for a project, and will be further discussed in Chapter 5. There are many benefits in being certain what vendors are required for the development of your project, and performing a smart and complete procurement and selection process for your team will allow you to make a conclusive decision on who is to become part of the project team.

Let it Flow; It's the Right Team!

Having the confidence that you have gathered the best possible team members, gives stakeholders the confidence that their vision can be put into action and turn into a reality. Even if the team members have come together for the very first time to work on your project, the team selection has been made, and it is time to get to work. Your duty as a project manager is to become familiarized with everybody's contractual responsibilities. There will be several meetings to allow for the team to get to know each other and to become familiar with your project management style, and to explain what you already know is important to publicly share about the project.

It is your duty that every single team member understands the mission and specific goal of the project. More specifically, the owner's definition for success must be shared and discussed with every team member. Any governing rules and regulations, or special conditions, are to be reminded. The stakeholder's directory is to be updated and shared with the team, and communication protocols must be established. It is extremely important that you, as the elected and appointed project manager, become the point of contact for everybody during project development. You must always be included and notified in all project communication and decisions.

Once important information has been disseminated, the team is to begin working its magic. Make a decision when to micromanage any portion of the project, and when the opportunity presents itself, take action to send a clear message that you are paying close attention and are knowledgeable of the team's processes and responsibilities. Begin meeting individually with team members working in active phase and daily tasks, and keep things moving along, always participating in finding solutions to any arising challenges or opportunities to make important decisions. Recognize great talent immediately, and praise daily successes. Allow the team to know you respect them; when they impress you with their work, verbally recognize them.

It is important to be flowing with the stream instead of acting as an obstacle—questioning, doubting, and rejecting ideas or solutions made by the team. An experienced, seasoned project manager trusts his own decision-making process, and can remind himself that he or she has already chosen this particular team for a reason. So, let things flow, and be confident that you have the right team for your project.

Chapter 5
Efficient Ways to Get Your Project Done

What is "Efficient?"

The art of achieving maximum productivity, with minimum wasted effort or expense, is considered to be *efficient*. The word, *wasted*, jumps out of the definition, and rightfully so, since wasting anything is the opposite of being efficient. Secondly, after you have reduced or eliminated waste, you must concentrate on maximally producing or realizing any task. As you learn to do those two things, and incorporate those key concepts, you are on your way to becoming an efficient project manager who works in a well-organized and competent way.

In project management, being efficient is directly related to your actual knowledge and planning abilities, and your capabilities to effectively execute your plan. Efficiency is also related to saving time, or doing something in less time than the actual estimated or usual time assigned to a specific task. Planning according to your extensive knowledge and experience, and guiding a project per the approved plan, maximizes the potential for the project, and its development process for being efficient.

One simple technique you may use to increase efficiency is to always be two steps or more ahead of your own schedule. For instance, you know that a payment application will be submitted for your review, processing, and formal approval for payment. Let's say you are in design phase, and are working with the architect of record. In your initial meeting, you may require him or her to come prepared with a tentative schedule for their portion of the work, and use that to agree on his monthly invoices and specific payment amounts. As long as the architect stays on schedule and sends you the payment application on the agreed amount and completed services, all you have to do is just sign it. This is a win-win situation since the architect can also save time and let an office employee handle his invoices way ahead of time. As a result, the architect will get paid on time, or even ahead of time, just because you decided to move an important task to the beginning of the process while working with this vendor. You, as the project manager, along with the architect, the financial department from the architect's office, and the project owner, all have saved time and energy, and that is what being efficient is all about.

Before, during, and even after project development, always ask yourself: "How can I be more efficient?" Any and all ideas that come to mind, simply try them out; if they are successful, choose to make them part of your process on all your future projects.

Organization Based on Efficiency and FLOW

The amount of people and tasks required to properly and successfully complete a complex project are immense. Subsequently, intending to keep track of them all in your mind is not an efficient way to get your project done. A single organizational chart, depicting names and assigned tasks below stakeholder's titles, properly organized by project phases, allows for an efficient way to channel information for proper review and approval.

Challenge yourself to become a project manager who is a master at organizing people and tasks, and understanding the efficient flow of the project. The stakeholder's registry, commonly known as the entire project's directory, is such an important document to keep updated for the benefit of effective project management. Relying on a complete organizational chart is a clear way to see the entire inception, review, and approval process for each and all of the project phases.

Organizational charts, with detailed information regarding the review and approval process, can show the flow of important information, either up or down. Up information may include items such as change order requests, payment application requests, or allowance use authorizations. Down flow information may include an owner's requested change in scope, or failed inspection reports and a corrective action plan. If an owner saw graphically, in an organizational chart, how many people he is affecting with his *simple request to change scope,* he or she may reconsider his request.

So, use it as a tool to make people aware of the importance of working as a team, and to help team members before and after your position in the organizational chart, especially when those who come after you are done with your portion of the project. Doing the best possible job allows for flow to continue; thus, promoting continuity and efficiency.

When I train teams so they may advance their knowledge, and polish their management skills, I concentrate on assisting them to polish their organizational skills. To schedule a introductory consultation, please visit **www.thebookonpm.com.**

The Knowledgeable Team Effect

The use of knowledge in a project is a strategic ability you get to utilize as an effective project manager in order to lead the project to a successful completion. Hence, you ought to define which areas each vendor or team member is extremely knowledgeable in, and allow them to simply share their knowledge for the benefit of the project.

In all cases, when your project has the privilege to have an extremely knowledgeable team, your project management abilities will drastically improve. Your management judgment must be put into action to allow for secure flow, without the need for micromanaging your expert team processes, which in most cases may slow the progress. This may become your greatest challenge; therefore, it is important to recognize the team's acquired and apparent knowledge, and to confidently step aside and trust them to carry the project on your behalf while you remain present and learn.

The apparent confidence of a team, and the accuracy and completeness of any task in any project phase, is the result of their application of knowledge. This is the perfect opportunity to promote open communication amongst team members. It is the actual sharing of the correct information that promotes and leads efficient execution of a task. Lastly, knowledge that is put into practice leads to a meaningful experience.

The Experienced Team Effect

Proven talent and tested knowledge via actual completed successful projects is the main formula for an experienced team. Truthfully, working as a team with the best of the best, results in dramatic positive results, phase after phase, all the way to successful project completion.

The proper management of experienced teams is how new and great, useful things are invented, and amazing complicated projects are realized. Watching an expert at work is addicting. Often, most people around an expert feel that they, too, can perform such tasks, and are inspired by the perfection of the expert's work. Anytime you watch a sports figure perform—such as the case every time Michael Jordan played basketball—millions of people around the world played basketball immediately after watching a bull's game, and took on the game just because they were so inspired by his greatness. When you have an expert or a master handling anything on your behalf, you, as a project manager, will be inspired to elevate your own abilities and become better at handling project management.

Recognize when or in what phases of your project you may require an experienced team handling that portion of the project. Assigning the experts in the most critical phases of your project is an efficient use of time, energy, and financial resources. In fact, it may cost you more to hire an expert in those areas. However, having a team because of inexpensive costs, may lead to disastrous results and expensive fixes, wasting time, energy, and money.

Having an experienced team will certainly lead to great results, while managing scope, budget, and schedule. Peace of mind throughout will also be experienced at maximum capacity.

Recognize the Actual Status of Your Project

You show up to the jobsite to assess the progress after the production team had a forty-hour week. You have your brilliant schedule in your hands to be used as a tool to gauge progress.

The team has made significant progress, so you are ready to write a site visit report from a great template, which has all the pertinent questions and plenty of room for your notes and comments. You take photos and video, and you head back to your office. Now, you are ready to recognize the actual status of the project.

As you create your progress summary report, make a commitment to be completely honest as you fully report on scope, budget, timeline, risks, quality, and a percentage complete, based on the overall schedule. Sometimes owners get feedback from other sources, and honesty will be your best practice if it comes down to verifying the validity of your report. Reporting on all areas of the project helps your efficiency and saves time as you avoid having several separate meetings.

If the project risk is high on any project areas, communicate it, along with potential solutions on how to mitigate or eliminate such risks. Any solutions you present, make sure you've previously discussed them with your team prior to presenting them to the project owner. It is important to report on key areas without inundating stakeholders with too much information, especially if it is technical information.

Staying in control of the project also allows stakeholders to sense and see that the project is in great hands, and that you are on top of things and two steps ahead. As a project manager, you are the project owner's representative, and your contractual obligations are with them; however, you are also representing the rest of the team members who are not there to make a case for how things are unfolding.

Reporting on the entire status of the project also allows you to exercise some stakeholder's accountability, and remind them of their responsibilities with approving documents on a timely manner, making payments on time, and respecting the team's rules that govern the project. Honesty is part of best practices, and learning to openly report on the challenges, as well as on the successes of the project, is the rightful, honest thing to do.

Who Was Before and Who Comes After

One of the most beneficial practices to promote efficiency in your team is to open the communication lines between a vendor, his predecessor, and his successor. Knowing who is *before* a specific vendor allows them to have important conversations on completeness and what is required before hand off. And when that vendor is done with his portion of the work, they can have the same conversations with his successor, and make sure they mutually agree on their hand off requirements.

This simple practice of connecting people to talk about the project allows for removal of misunderstandings or assumptions, and also serves as an opportunity to learn if scope is being properly addressed as completed. Different vendors may have different hand off requirements; therefore, setting a standardized easy-to-follow method may be beneficial. Anytime you are able to systemize a repetitive task, and the team becomes familiarized with it, you and the entire team will save time and energy.

In Chapter 6, you will be able to read about systematization, and learn some of the benefits. There is also sophisticated software, which can computerize some of your project tasks. Some of the available programs are pretty sophisticated, and they offer notification services customized by the requirements of the project's schedule. Do your own research, and use it only if it is proven to be reliable, and if it actually saves you time.

Chapter 6
The Genius Way to Organize Your team

Monotony = Disconnected Team

It is true that repetition makes you better at anything; however, during project development, a lack of variety and interest, paired with a tedious routine, may be affecting the moral and emotional wellbeing of your team. Anytime you have a team who is bored working on your project, you are close to having a team that is totally disconnected from your project. After disconnect, comes inefficiency, and inevitable mistakes or inaccuracy with scope compliance.

There are many ways you can avoid monotony, or promote variability, by allowing your team members to fully understand the project's mission, and not just the task at hand. In Chapter 9, you will read about the project's mission and its importance. Another way is by rewarding early finish and accuracy so they are excited about accelerating their portion of the work, and being accurate and in full compliance with scope.

You may also share some exciting news about the project, such as plans to submit for a worldwide recognition award, or being acknowledged on important social media sites as you recognize them for their portion of the work. The main idea is to avoid monotony, and make your team feel appreciated as you periodically remind them that their portion of the project is just as important and meaningful as the rest.

> *For other great ideas on how to keep your team motivated, please visit **www.thebookonpm.com**, and download a complete list of ideas and techniques, which will keep your team energized and connected to your project.*

Systematization Is the Mother of Progress

How do you create a system without falling into monotony? Simple: allow your system model to accommodate for individual expression. This means that as long as the team's creative choices, during execution, yields the established and required system's results, team members can decide how and in what ways their execution may be more fun, efficient, and variable for them.

Having a system based on the desired end results allows for creativity during the actual implementation process. You may create a system by being clear on the defined final product, and give them a not-to-exceed time per schedule; then, simply let the team work their magic. A flexible system may evolve and improve itself; however, it is important to establish the rules of the system and give your team constrains such as time and budget.

The goal of having a system is to aid organization and improve predictability. Furthermore, a system allows for new team members to enter the system and become productive in the least amount of time possible. Having phases in a project is part of the big picture of a system utilized to complete the entire project. In each phase, there may be several unique approaches executed by different team members; however, the goal remains to complete that specific phase based on the actual defined scope requirements. You, acting as a project manager, are also part of the system, and your own creativity is used to apply your acquired knowledge and skills for the benefit of each phase, and the entire project.

As you participate in the development of diverse projects, you are able to document key activities driving the project, and turn them into templates; this will save time.

 *To improve your system, you may download effective project management templates by visiting **www.thebookonpm.com**.*

As you become more knowledgeable in the process of improving your own management system, you automatically focus on allowing your specific system to be repeatable and scalable. When assigning work periods to your system, keep them to a 12-week maximum to bring all long-range goals into view. In fact, action plans are better executed if they only span from 30 to 90 days. This allows your team to stay motivated as they keep moving the project forward towards successful completion.

Expertise Is Married to Accuracy

Having expert skills and knowledge in a particular field drastically increases your accuracy when completing a specialized task. Your expertise in project management is what gives value to your daily work and your brand. Likewise, recognizing your team's individual expertise will allow you to properly delegate tasks and increase accuracy to the point of achieving perfection.

Working with a team, where the entire team is an expert in their field, is precisely how a spaceship can go to outer space and safely land on the moon. The moment you are able to recognize everybody's expertise, accuracy will govern, thus saving time and money. Although expertise comes with a price, it still makes sense to hire experts in the critical areas of your project. For instance, you do want to hire an expert on expansive clay soils to design your foundation system, and not just any structural engineer without the proper dedicated experience just because the requirement is still met and the cost is significantly less than the expert structural engineer.

As you seek for individuals with specific expertise to complement your team, make sure you verify that they are indeed experts. Usually, experts are very easy to spot once you realize that experts only focus on one specific subject; they have plenty of practice working in that specific area, and their end product is perfectly accurate.

To download a unique and effective checklist to help you identify when a vendor is truly an expert, please visit **www.thebookonpm.com**.

Focus Is Completion's Best Friend

You begin a brand new day with a detailed to-do list, and it includes a manageable and achievable list of tasks, with a few of those tasks labeled as critical to complete that day. You are ready to get to business to complete all tasks, and you receive an email from a local TV station, asking if you would like to be featured on a special live program to be interviewed and broadcast your unique project. You consider it and make a decision that the opportunity is beneficial to stakeholders and the community. Besides, it will help your brand and allow you to share some exciting news with your team. So, you do it, and this unplanned venture takes 4 hours of your day.

You realize that there are not enough hours in the day to complete all items on your to-do list. The fact is that you switched activities, and your shift in focus prevented you from completing some tasks deemed important. This example is just for one single day; losing focus on the main goals of the project is sometimes devastating to your schedule and budget. Keeping your entire organization connected to their responsibilities, and strictly focused on the phase and overall project's mission, is the key to a successful completion.

As a project manager, you must develop your ability to identify all distractions, which come in many different forms: sometimes as awards opportunities or free publicity, and other times via stakeholders wanting to impose new tasks not deemed important or required to the successful completion of the project. Managing your team so they pay particular attention to the important required tasks and goals of the project during production or execution is basically how staying focused will allow for a successful completion.

Confidence Leads You to Problem-Free Completion

In project management, confidence in your team and established processes arrives after the systems are defined and implemented, and are working flawlessly. Even when you are facing challenges, your established procedures to find and implement the solution are working perfectly; therefore, your confidence grows even more.

A problem-free project does not necessarily mean that you had no problems during project development; it means, instead, that you and your team were able to solve all encountered challenges. Therefore, the project is truly free of any troubles or complications.

Confidence arrives because of the many key decisions you made as a project manager in responsible charge of the project. If you are implementing many of the ideas shared in this book, and you are experiencing positive results, your confidence in your abilities is rightfully growing. When you opted to select the best team your budget can afford, and have placed the right people in the right places, and your system is flexible and allows for solving of any issues, you can confidently say that you have set up the project to be completed without any foreseeable problems.

Cleverness Is the Basis for Amazing Unstoppable Teams

Ingenuity in your organization sets you apart and rightfully leads to amazing, unstoppable teams—often, unique teams accomplishing perfection unlike any other. Allowing your team to be clever in their process of execution invites flexibility, creativity, and freedom to any project development. Clever teams are healthy, happy, and adventurous; they are very alive and notably confident about their skills and knowledge. Clever teams' main focus is finding new ways to use ingenuity as they efficiently get from point "A" to point "B" but with a great sense of discovery.

As a project manager working closely with stakeholders during procurement and selecting teams with a defined diversity, find the best areas where these clever teams are able to have the greatest impact, and can yield the best results for the benefit of the project. Clever teams must also have a defined purpose and an expertise; your duty, as a project manager, is to inquire and learn what their proficiency and capabilities are; then, simply place them in the right phase of your project.

When I work with organizations seeking to maximize their potential, I help them identify new ways to promote the ingenuity of their teams. It is a revealing process to experience, and the results always surprise the entire organization.

Clever teams are amazing problem solvers, and they are also very creative. Their diverse way of executing anything creates a positive, fun energy environment. Additionally, core values of a clever team may include enthusiasm, perseverance, curiosity, enjoyment, openness, contribution, support, and creation. As you plan and work in the different phases of the project, the core values of a clever team may help you decide where they may best fit and complement the team's efforts. Let cleverness assist your efforts, as a PM, to become an unstoppable team.

*To hire an experienced project manager to maximize the productivuty of your clever team, please visit **www.thebookonpm.com** to learn how your company can receive the proper guidance to maximize efficiency and increase your profits.*

CHAPTER 7
THE RIGHT WAY TO SCHEDULE ACTIVITIES

Elements of a Schedule

A schedule can basically be considered a complete guide comprised of the entire scope of the project, expanding the full term from beginning or notice to proceed, or approval of project charter to the very end, or the agreed last contractual date. In other words, a schedule is basically the *WHAT needs to done, by WHEN,* and *WHO is responsible for it*. The project schedule is also required to comply with contractual milestones, and it must be formulated with the level of detail required by the contract. Simple schedules basically include the lists of tasks organized by phases of the project, and the tasks' initiation and completion dates, also known as the tasks' duration.

Schedules can serve as a daily, weekly, and monthly guide, and they are a reliable tool to verify *estimated vs. actual progress* for the completed portion of the project. The level of detail or task description in a schedule varies depending on whom the schedule is written for, and what the true intent is for having it. A schedule given to the project owner during procurement, in response to an owner's inquiry regarding vendors understanding of the project's *potential development schedule*, will lack many of the real aspects and crucial information of a usable schedule during the actual production phase of the project.

An important element of a schedule is relationship and dependency of diverse tasks. Informative and more detailed schedules may also include material procurement and arrival on site for the major critical tasks. You may smartly make a ruling and request that all the items with costs over 500k are to be inclusive of material procurement and required manpower, and be clearly defined on the schedule. If specialty equipment is to be utilized for key specific phases, it must also be listed on the schedule.

In sum, schedules must serve as a guideline or a checklist of things for project managers or supervisors, in order to guide the project and realistically verify the status of the project as completed.

New Ingredients to the Schedule Plate

Taking schedules further to serve a greater purpose, you may consider adding enough information to allow listed tasks to act like a to-do list. In fact, including critical reviews and approvals, and listing inspector's information, may be extremely practical and beneficial. Adding the sequence for review and approval, and listing appointed personnel to perform such duties, will be a great way to organize and make the entire team aware of critical inspections events.

Enlisting if there is any available float to alleviate some of the tasks while working under pressure may also be useful. Regarding manpower, listing when record numbers for manpower will be on site, along with the allocated duration in hours, is helpful. Doing so is highly beneficial, especially for project owners wanting to visit the project. Since you know ahead of time that manpower on site is high, you may note on the schedule—NO VISITORS ALLOWED—during that period.

Every time you catch yourself thinking, "I need to remind myself to do this," and that specific task is not on the schedule, simply add it, and test if having such information listed is beneficial to the entire team.

 *To schedule an introductory meeting to review your project schedule, and have the expert advice of a project manager, please visit **www.thebookonpm.com**.*

Look Further into Critical Tasks

Is it possible to dig deeper into critical tasks and find information worth sharing on the project schedule? Absolutely yes! If you further investigate and understand who is at the bottom of any task, performing any type of work, including review or approval, then adding their information to the schedule will provide you with another level of management.

For instance, on your schedule task labeled, *"Submit to government entity for building permit"*, and you allocate the known 6 weeks for plan review, comments, and corrections, you are not actually looking all the way to the bottom of that specific activity. Therefore, your task is to include what reviews will take place, and names of personnel performing each review per each of the required disciplines. If the reviewer of certain discipline has specialized sub-reviewers or specialty procedures, list those as well. If the plans are on hold, or not moving forward, you have the required information to catch that and contact the right personnel to inquire about moving it forward.

Looking further into critical tasks is particularly beneficial on all critical tasks taking much of your schedule time, and being performed by stakeholders whose time you do not get to directly manage. Although many of these tasks are outside your immediate control or supervision, knowing their process can help you and your team properly manage them; in many instances, you may learn about a better way to handle that specific task.

On a schedule-driven project, during project planning, two items were identified as potentially delaying the schedule, and potentially missing the small window allowed for completion of the project. The project manager took the lead and made a decision to handle those items first, and prior to formally procuring for the construction team. One of the items was an environmental/health requirement on asbestos abatement and demolition; this task required special testing and special permitting, and the need for engaging a separate specialty contractor. The second item potentially delaying the schedule was ordering and receiving customized HVAC equipment that would normally take longer to buy and ship than the entire schedule for the project. Both items were handled directly by the project manager on behalf of the owner; when the construction team was hired, scheduling and completing the project as originally intended by the project owner was executed ahead of schedule.

Factors Shaping Your Schedule

There are other factors shaping or affecting a project's schedule, and knowing these factors exist helps the team to plan and allocate time just in case they arrive. Although some of these factors impacting your project schedule, such as climate and accidents, are completely outside your control, you may still formulate a plan to mitigate their effects on your schedule. In the case of climate, knowing climate trends and identifying rainy seasons can help you plan for such an event when they are aligned with your schedule activities. Perhaps activities for the project, during those days, happen indoors, or other critical tasks detached from climate can also be advanced.

Other factors difficult to plan or accommodate in your schedule include sick people, new hires, new trends, changes in scope, additional safety, omissions, new regulations, unforeseen conditions, etc. You, as the project manager working with your team, may formulate and agree on a plan of action to take if any of the previous factors happen to arrive on your project. Having an agreed plan of attack drastically reduces downtime and excuses for not having a potential solution in place.

How to Address Unforeseen Conditions

In every project, unforeseen conditions and events arrive and test your skills and ability to find solutions. There are some basic steps you must take in order to address unforeseen conditions. First, immediately attend to it by identifying who is in responsible charge of that specific portion of the project.

Secondly, once you identify the appropriate party, perform an internal 911 or urgent phone call to the individual in responsible charge, and explain the emergency and the urgency for them to address the issue at hand immediately. Third, look at your schedule for float, and identify items you may borrow time from to address the unforeseen condition; then, take necessary measures to resolve the issue effectively. Lastly, choose to plan ahead of time for unforeseen conditions by creating an allowance in your budget for such situations. One of the main reasons unforeseen conditions do not get attended to immediately is because there are no funds allocated to pay for the required resources to solve the matter at hand.

Although unforeseen conditions are unanticipated or unexpected circumstances or situations affecting the final price and completion time of a contract or project, they are not omissions or mistakes made by the team. There is a fine line between unforeseen and omissions, and the best way to know if any condition qualifies as unforeseen is by verifying that the situation was not part of anybody's contract. For example, if while digging for a new building's foundation, you uncovered buried fuel lines attached to large tanks buried underground, and they now need to be properly removed, and new dirt needs to be imported to fill the void, this would be considered as unforeseen, but only if nobody in the team had it in their contract to investigate underground conditions, including to do an in-depth analysis of the history of the project. If you also paid somebody to investigate and perform underground investigation, this would be very difficult to sell as an unforeseen, since buried metal fuel lines are easily detectable.

The proper review and verification of unforeseen conditions as they appear is best performed by you, as the project manager, simply because you know the entire history of the project, and you understand all the team's contracts and their content. Follow the steps listed above on every project you manage, and you will certainly handle any unforeseen conditions as a true professional in responsible charge, managing the entire project development.

✚ *To benefit from the expert advice of a project manager, please visit **www.thebookonpm.com** to receive assistance reviewing your project's documents, to identify potential unforeseen conditions prior to officially starting your project.*

A Brilliant Kick-Ass Schedule

By simply following all the information outlined in this chapter, and with the use of reliable scheduling software, you, as an experienced project manager, may create a brilliant project schedule.

The best way to test the effectiveness of your schedule is by putting it to a real test; so, have it used by a different PM in your organization who has no previous knowledge of the project, and see what happens. If they are able to take it and run with it by simply following the information on your schedule, then you have created a *Kick-Ass Schedule* that any project manager without previous knowledge of your project is able to understand and manage, at any time and during any phase, all the way to completion.

CHAPTER 8
A CHALLENGING PROJECT STORY

What Creates Unwanted Challenges in a Project

Inexperience, unfamiliarity, and simply not knowing the required steps in order to take a project from point "A" to point "B", while verifying it to be in full compliance with the intended project goals, will always create unwanted challenges. These unwanted challenges or problems will appear throughout the entire development process; however, they typically multiply during the production phase. If you are ignoring critical aspects of a project and somehow manage to get it to the end, during the close out phase, you will face your greatest challenges yet. At the end of a project, any successes, as well as all challenges, will appear and arrive to remind you that things have not been going well, and to hope for a clean project close out may be nearly impossible to accomplish. In any case, your upcoming meeting titled, *Lessons Learned*, will be very extensive and informative, and it will be filled with many instructions on what not to do ever again while you are managing a project.

A project, which was approved with a specific scope but without the formal review of an expert, will most likely lead to the wrong procurement and final team selection, which will then lead to gaps during production; and, at the very end, it will result in unforeseen, expensive conditions. Mistakes made at the beginning of the process have a way of showing up throughout the entire development process.

Below is a list of some activities or practices that may cause unwanted challenges, and if the listed activities are properly handled during the planning stages, you will be creating an armoire, or a protection shield to repel unwanted challenges.

- Unrealistic Owner Expectations with Scope, Budget, or Schedule
- Inadequate Funding
- Inexperienced Owner or Stakeholders Making Critical Decisions
- Procuring and Selecting a Team Based on Costs Instead of Experience or Expertise
- Insufficient Contractual Responsibilities
- Lack or Inability to Exercise Stakeholder Accountability
- Stakeholders managing the project without the proper credentials or experience

Something as basic as creating a comprehensive list of stakeholders is an ability that you, as a project manager, must develop. Knowing the process, and having the required experience managing projects, allows you to identify absolutely everyone involved in your project who may influence the outcome. Certainly, you are in a better position knowing members from the cleaning crew to the secretary processing your request in government agencies, to the board member president signing approvals, to the IT quality inspector, and the fire department chief and inspectors—even the temporary replacements for key and very important team members, who may leave or be absent during a critical phase due to illness or any other reason. So, the answer is yes! Everybody must be in your stakeholder directory. And yes, even that person you are

A Challenging Project Story

also thinking of.

On the bright side, if you ever manage a project with many challenges, your ability to solve problems, and the fact that you are able to carry on, will probably be the perfect opportunity to develop your problem-solving ability to its full potential. Solving unforeseen conditions and moving forward is a great ability that you as project manager must advance. An even more important ability you ought to develop is your proficiency to identify and mitigate potential challenges, even before they arrive. Develop this ability, and you will be rightfully considered an expert in your area.

> *When I work with project developers or project owners, we work as team to eliminate all practices that may create unwanted challenges in their projects. This professional partnership translates to effective use of time, money, and resources.*

The Crisis Manager

A project just came to a halt. Inspections have not passed, and several project documents are not in line with code requirements. Furthermore, proper permits from government agencies were not properly secured, and the project owner is extremely frustrated. Payment applications do not match the progress as invoiced, and the general contractor has issued a stop work notice for lack of payment. Suddenly, you receive a phone call from your office, and they inform you that you are assigned to this project effective immediately. You are now the official *Crisis Project Manager*, and you have taken the responsibility to jump into a project that is clearly in trouble.

If you are ever invited to handle or are assigned a project in crisis, it is a clear sign that you have gained the respect and trust from the project owner, sponsor, or your employer. If you are a great experienced project manager, you probably have gained the respect and trust from all three entities already. Project managers, who can see the big picture and understand every phase and every single detail for the complete development of a project, can in fact, and will certainly, jump into the opportunity and take over a project in crisis. It is like a doctor performing surgery on patients diagnosed as inoperable or incurable. The doctor goes in with all he has, to do what he does best, and cures a patient deemed incurable.

But what is considered a crisis in a project? It's simple: when the project gets chaotic, wild, and simply out of control, and there is absolutely nobody in the team able to work with all the arising challenges, doubts, questions, inquiries, accusations, etc.; and when vendors begin taking advantage, as they notice nobody is watching over the project; and when vendors realize that nobody understands contractual obligations. Vendors or project contractors begin quietly deviating from the agreed scope, and the changes may be subtle; however, as they gain confidence that nobody is noticing, a few things may be completely taken out of your project. If you are a naïve project owner, or inexperienced sponsor without a project manager, you will never be able to know the difference when production is not in full compliance with approved scope. Crisis in projects may be compared with automobile accidents. Drivers never notice what is happening all around them until the accident happens, and all of a sudden they are in panic or in a serious situation in search of a lawyer to help them make sense of insurance claims, police reports, traffic conditions, automobile manufacturer potential omissions—which may have caused your accident—and, of course, somebody to write up the lawsuit and hopefully find blame in others to prove your innocence. This does not mean that you need to drive with your attorney in your car just in case an accident happens. It only means that there are expert professionals to address special scenarios and circumstances; in case of an accident, you, as the driver, require expert advice to represent you by handling everything you as a vehicle owner and driver do not understand.

> *To benefit from the services of an experienced project manager, and to get your project out of a crisis, please visit* ***www.thebookonpm.com*** *to learn how the proper management techniques will securely get your project over the hump.*

You, as a project manager, once assigned to handle any project in crisis, must assume the responsibility to solve any and all project issues and challenges without finding blame. You must diligently work towards finding a solution, without the need to blame team members or even comment on how incorrectly processes were implemented or tasks were handled. It is the professional and right thing to do. Let attorneys do their job, if and when their services are engaged. Managing a project in crisis is about finding solutions and implementing the best solutions to gain control of the project with a clear mission to get it back on track. Your immediate goal ought to be to do whatever it takes to guide it and manage it closely to a successful completion.

When you have the opportunity to act as a crisis manager, your stock just went up, and it is a perfect opportunity to display your knowledge, expertise, and acquired specialized management abilities. If, on the other hand, you are a project owner or a sponsor, keep in mind that one sure way to get your entire team and your project into a crisis is to choose to develop a project without the services of an experienced project manager.

Watching Conflict of Interest Closely

Perhaps you already feel that you have way too much on your plate as a project manager. You are working in all different phases of the project: from initiating to planning, to execution, to monitoring, and lastly, to closing the project out. Nevertheless, it is extremely important to keep a close eye on any stakeholder's personal interest not clashing with professional or public interest. It is a fact, for some people, that the saying, "Rules are meant to be broken," holds true; so, they feel they can bend or break the rules, thus creating a conflict. This practice most likely will affect scope, schedule, or budget, as well as the morale and reputation of the entire team and organization. It is an unpleasant, uncomfortable, and unethical when stakeholders put themselves in a conflict of interest scenario. When this happens, stakeholders are very unreliable, and it becomes exceedingly difficult to trust them to be part of your team.

On many occasions, an inexperienced stakeholder or project sponsor will be handling certain situations in which it is apparent to you and everybody else that they have vested interest in money, status, or reputation, or all three of them. However, you, as the project manager, may not realize that they genuinely ignore the fact that they are actually violating the rules and, in the case of public projects, that they are breaking the law. One way you may know they completely ignore that they are creating conflict, without even knowing, is by the simple fact that they are open about it during official meetings, and even document their unlawful practices officially via email or by any other written means. Project owners or sponsors, who are placed in powerful decision-making positions, many times lack the political experience and business sense, and completely ignore the existence of rules and regulations. They may even believe they are doing a good deed by helping a family member grow

A Challenging Project Story

their business. These types of arrangements not only are illegal, they also are not desired as part of your project practices. Having team members who feel they deserve and are entitled to special treatment, and that they are untouchable, creates an environment difficult to manage, and difficult to keep in line—certainly within the legal boundaries.

Therefore, you must establish a procedure, and rely on expert advice, while investigating what is considered conflict of interest in your project. Have the legal and the human resources, or the appointed department, determine what an acceptable practice is, and what is considered conflict of interest. Perform this duty before procurement and signing of contracts for all team members. As you create your stakeholder directory, watch for potential conflict of interest, and immediately bring to the attention of the appropriate party, any potential for it to take place. Again, you, as the project manager in responsible charge, must be aware of absolutely everything that is going on with your project. Unquestionably, you are entitled to step aside if you feel unethical behavior is taking place, causing a conflict of interest, and the situation is beyond your control. If you choose to stay as a project manager, solve this issue, and prevent it from happening ever again. Establish protocols so all stakeholders understand that you are watching every aspect of the project. Remind them as required that you are meticulously working towards having a transparent, legal, and honest development process, without any conflict of interest.

During the development of a certain public project, the project sponsor was involved in many phases of the project, mainly to learn about all intricacies of a project moving from phase to phase. During the furniture, fixture, and equipment selection process, the sponsor requested to participate during review of all furniture selection, and requested to meet with the potential vendor who was being considered for the project. After the meeting ended, the project sponsor requested to hold another brief meeting without the presence of the project manager. The following morning, an email arrived for the project manager, stating the good news that the sponsor had worked out a great deal with the furniture vendor, and that he had successfully negotiated a 5% donation back to a public entity to aid the organization's business expenses, such as employees' salaries and bonuses. The total FFE contract was budgeted at $1.8 million; therefore, the inappropriate donation would have been for $90,000 from this particular vendor. The project manager proceeded to forward the email to procurement and the director for business services of that public organization. After a meeting with the legal department, and verification of statutes, it was decided that the vendor and the project sponsor were in a clear conflict of interest, as the sponsor held the power to approve any contract amount which, in return, would directly benefit the public organization the same sponsor was serving. The vendor was officially dismissed and prevented from being able to participate in a new formal public bid, where the project was officially and publically awarded to a different furniture company. The awarded company had no previous contact or any type of relation to the sponsor.

After many meetings with the appropriate team members, the project manager was able to conclude that the project sponsor ignored all these rules, and that he truly was unaware he was in major violation of several governing laws. Fortunately, for the entire team, this was caught in time, and the furniture project was awarded to a completely different vendor. The procurement, purchase, delivery, and installation went smoothly, thanks to the watchful eye of a project manager fully immersed in every move of the project.

A Challenging Project Story

The potential for a similar activity happening in your project is very real, and it is your duty to mitigate or eliminate any activities that may lead your project into a conflict of interest battle. Stay alert, and always do the right thing, and your reputation will remain intact.

A Project Going South

During the solar energy boom, many public projects made a conscious decision to *Go Green,* and began searching for the perfect solar energy partner. It made perfect sense since most public buildings, such as public schools and other government buildings, are occupied and operate during the day and under extreme sunlight; at least that is the case in the Southwestern United States. So, that is exactly what happened when public school districts organized themselves, and procured for a vendor to provide solar energy production via solar panels and sophisticated monitoring equipment in their campuses. The scope was defined, the vendor was procured, the proposal and contracts were presented, and the opportunity was unbelievable and impossible to turn away. The solar vendor offered free equipment and installation at no cost to school districts; in some cases, the cost for solar panels and equipment, including installation, was in the millions. For an 80,000 square foot campus, producing about 30% of their energy requirements via the sun, the cost of all equipment was in the millions. The decision makers and stakeholders in responsible charge to approve such a contract jumped at the opportunity to *Go Green,* to produce solar energy, and pay $0.00 for the equipment and installation. So, they did.

"Nothing is free" is a famous quote, which is 100% applicable to this unbelievable offer. There are not enough pages in this book to talk about the many mistakes stakeholders made; and

just in case you are wondering, yes, a project manager was present during the negotiations and agreement of the solar energy contracts. The short version is, what the public school districts hoped and assumed they were getting was far from what they signed for in their solar energy contract. After a few years of producing solar energy on their campuses, they realized that it was costing more: they were paying, in some cases, about $55,000 more per campus in their yearly energy bills. As soon as they became aware, it was time to go back in time and learn what had happened. It was also time to find solutions to get out this awful deal. It was time to bring the attorneys to make sense of this mess. So, they did.

This solar project is a perfect example of how mistakes made early in the process, during scope definition, vendor procurement, and negotiations, can and will cost you time, money, and resources throughout the life of the project. It also proves that inexperienced project managers handling projects is even worse than not having a project manager. Therefore, as a project manager, if you are ever requested to handle a project of which you have no past experience, be honest about it—do the right thing, and procure and hire the services of expert advisors.

If the project manager for this solar project, which needed to face south in order to maximize solar production, had been more experienced and knowledgeable, or simply honest about his limitations, this project would have never gone south and, as a result, millions of dollars would have been saved and utilized for the benefit of students in public education.

Am I to Save, Fix or Pull the Plug?

There may be unwanted challenges, and your project may enter into a crisis as you realize potential conflict of interest is in your project—which, by the way, may be going south in a hurry. What do you do? Do you save it and fix it? Or do you make a tough decision and pull the plug?

As you expand on your knowledge and gain more experience, you will begin to realize that project management is basically about completing things right, fixing things when they need to be fixed, and eventually successfully finishing that project. It is never the intent of a sponsor or project owner to bring you as their project manager just so you can suggest and strongly recommend terminating everybody's contracts, and end a project. Even when the project is severely suffering, if you can visualize a way to bring it back in line, and see a clear way to salvage it, simply choose to save it and complete it.

There are instances in which a project's only option is to terminate it. The most obvious reason is when financial resources have been exhausted, and there is no money to save, fix, and complete a project. Not having the required funds or the financial means to complete any project is a strong reason why the project then must come to an end. Another not so obvious reason is that anytime a project is affecting and endangering public safety, you must also consider ending such a project.

*To benefit from the services of an experienced project manager, and assess if your project is salvageable, please visit **www.thebookonpm.com**, and schedule a professional consultation.*

As you visualize a way to save and fix a project, make sure you are considering important vital aspects of the project, such as availability of financial resources, public safety, code compliance, community, and potential detrimental environmental impact.

Spending money to fix a problem that can't be properly fixed will drain your budget and put your project in financial hardship. Challenging projects do exist, as well as impossible projects. Your experience as a project manager will dictate what project you label as challenging, and which one is simply labeled as impossible to complete—and now you know which ones are worth saving and fixing.

Chapter 9
Success Stories in Project Management

Definition of a Successful Project

Is an ecstatically pleased project owner enough reason to consider the project a success? Of course, it's not. Then, what is an all-inclusive definition for a successful project? Success is simply defined as *"the accomplishment of an aim or purpose."* Focusing on the obvious, if you hit your targets as defined in your scope, and the project is in absolute compliance with your budget and schedule, then you may correctly state that you have completed a successful project.

There are many different ways to measure a project. If you choose to measure it on the success scale, you will quickly realize that even success has different levels. In Chapter ten, you will read about how to assess your project results, and you will learn how to create your own measuring stick.

Success encompasses other not so obvious factors, such as *individual satisfaction,* as in the case when you focus your management efforts in discovering new opportunities to be more effective and efficient. Additionally, discovering or inventing new ways to improve the process of development is a great opportunity to celebrate and to feel accomplished or successful.

If a project you are managing presents an opportunity for you and your team to discover new techniques and methodologies, and as a result, new information is revealed, and as a result, it shapes the way projects are developed in your industry, how about that as part of the formula for success?

Success is also a matter of difference in opinions. What may be successful for your project may be considered just a basic baseline for other projects. Hence, knowing precisely what makes the perfect formula for success in your own project is a must. In defining the success formula and its factors, consider having success values, such as honesty, confidence, perseverance, integrity, innovation, and adaptability. All these values are assets acting as a solid base where success can rest upon. Celebrating your victories and embracing your challenges to learn and become better are also crucial factors to be included in your ever-changing formula for success.

Write your own definition of success, and share it with the rest of the project's team. It will provide a clear target and allow for recognition and celebration when the team aims and hits the bull's eye, and the project's purpose is accomplished. Watch for the team and stakeholders saying, "WOW," and use that as an indicator that success is surely on its way.

The WOW Factor Everywhere

In project management, what stakeholders and people in your team say matters. Words do mean something, especially when they come as feedback. The energy that is created when positive feedback is spread throughout the entire team is extremely powerful. Team members become more connected to the needs and goals of the project, and stakeholders appreciate representing a project that is the talk of the town, is on TV, in the newspapers, and on social media, portrayed as a never seen, never done, or simply as a project outside the norm and never seen or done before.

A great project begins its greatness the very same day it is envisioned; its sensation of importance grows the moment it's verbally described, and its distinctive goals are cleverly defined. Goals that are ambitious and considered challenging to attain usually trigger people to say things such as, "WOW, that is impressive." Since the project's inception began impressing people, it will be easier to carry that energy into procurement and get the entire design and production team. Every team member will be easily excited about working in a unique, challenging project that will bring up their brand.

Being different is often all it takes for a project to be recognized as a uniquely amazing project. It could be that the project owner hired you as a project manager for a project with a unique scope or a challenging budget, or perhaps an aggressive schedule. It may be a project that will have a huge impact on the community, or will allow users to accomplish something not yet accomplished before by anybody. For some time, projects getting a LEED certification were the thing to do, as every LEED certified project was considered as the friendliest buildings to the environment. The more projects that got their LEED certification, the less they made it on the news or were considered as a unique project to develop.

Project developers who understand the physiology of new development always look for a niche or a unique way to present and sell their project to the end user. Once you are in a project that has this characteristic attached to it, it becomes your responsibility to carry on the same level of interest, and deliver a project that can inspire everybody to be pleasantly impressed, and often say, "WOW, that is a unique project." Make it your own personal goal to hear the same everywhere, and the wow factor will carry your project to a successful, impressive completion.

To create WOW experiences on all your projects, please visit ***www.thebookonpm.com*** *to learn how the proper management techniques will create the WOW factor everywhere.*

Project Delivery with a Mission

Knowing the specific purpose for developing a project allows project management to revolve around a clear mission. A project manager who is aware of the main project's mission, and understands the "What," Why," and "Who," can easily focus all its resources around these reasons, and in alignment with their significance to the world. Following a project mission that is formulated correctly gives project members the desired focus to work toward a common, meaningful goal. Understanding the "Why" is the most important element of your project's mission. As a project manager, once you know why, you are able to direct your team's efforts, always respecting and revolving around the main purpose or mission. Moreover, you will also be able to identify when stakeholders are deviating from the main mission, and gracefully guide them back to stay the course.

Make it part of your own personal practice to ask the project owner or sponsor these questions: Why this project? And, what is the mission or main purpose for developing this project? When you meet the requirements in response to the "Why" of the project, feeling successful is inevitable. Your duties as a project manager were clearly defined in Chapter three, and now that you are aware of how a project's mission can guide your efforts, you can take on this duty with great pride and intent.

Project management allows for diverse goals and aspirations to be equally as important. It is implied that scope, budget, and schedule are your main obvious responsibilities and, for those, you are expected to use your knowledge and skills, which makes you the experienced project manager that the project owner decided to hire. Likewise, you, as the official project manager, are entitled to know the values and traits that define the development group. For skyscraper developers, their main mission may be to develop the tallest building in that city, or even in the world. Therefore, when meetings revolve about ways to save money, you already know that eliminating floors, resulting in reducing the overall height of the building, is not worth mentioning. Your efforts are best employed to revolve around proposing other strategies.

 When I am hired to work with any organization, together as a team, we dive deep into the true needs of the project, and clearly define a powerful mission.

As you expand your interest in understanding the clear mission of the project, you allow yourself to develop a more intimate project management experience with all stakeholders. Your senses are invited to get tunnel vision on what is truly important, and your efforts are efficiently utilized. As a result, you are certainly becoming an efficient project manager.

When the Goal is Clear – an NFL Project

For many years, the oldest school district in the State of Arizona envisioned providing an athletics complex for their students, to enhance the district's sports program and to aid their physical education goals. One of their campuses had the required real estate to add some playfields and, by doing so, eliminating an oversized open area, costing them thousands of dollars per month just in irrigation costs. The project goals were clear since day one, and the project manager had the privilege to be engaged early when the stakeholders began their brainstorming sessions. The project's mission was clearly defined and, as a result, the project scope was ambitious, unique, and mission driven. In a short time, the stakeholders and project sponsor reached the decision to build a NFL football field, with a world-class, 9-lane track around it, and a softball field right next it. Their scope strictly stated that irrigation was to be drastically reduced or eliminated, and that funding for the project was to come from at least four different sources, the main source being derived from *adjacent ways* funding.

The mission was so clear that the project manager used it as a checklist, and as the formula for success. The initial team members were organized, and the project manager worked closely with the assistant superintendent for the district, and the business director, to identify and pursue the diverse funding sources. Efforts were so focused on the clear goals of the project, and having the entire organization working as a team paid off in a big way. Their secured funding sources included a NFL grant, strategic adjacent ways funding, bond dollars, and capital funds strictly pointed at sports programs. The enthusiasm and energy of success for this project was able to be carried throughout procurement, construction, and close out phase.

The project was built, and the apparent accomplishments and realizations could not be ignored nor denied. First, it would be the only time that the NFL, through the Arizona Cardinal Organization, would award their grant to an elementary school. They also offered more money if the project would choose artificial turf and eliminate irrigation requirements, which was also part of the project owner's defined scope. The design team was so focused on the funding sources requirements that they were able to design the fields to be funded mainly from adjacent ways resources. The project manager proposed a design/build delivery method, which resulted in significant savings; thus, eliminating the need to tap into capital funds, and minimizing the usage of bond dollars.

The impact this project had on the students and the surrounding community was profound. During inauguration day, Arizona Cardinals players spent a day playing and training with district students, teachers, and community members. All of the envisioned and established project goals and aspirations were met as defined in the project's mission statement. The success of this project was carried throughout the entire development as everybody made a conscious decision to work as team. Clear goals painted a clear target. The project manager remained, managing the *district's project,* for over 6 years, spending close to 50 million dollars in project development. Every project had its own clear mission and great success story.

A Dream Is Realized – MIT

Dreams lead to visions, and when your vision is powerful enough, it becomes a reality. In project development fulfilling a need, which will revolutionize or change the perception of things and, as a result, shift the paradigm of an outdated tradition or way of being, a great project is in the making. Project management is best employed in projects where the bar is raised so high, it scares most of the organization members. Likewise, any dream worth pursuing is certainly a dream that shakes your whole being in a way that your very own fear is enhanced to new levels. This emotional state of being, feeling, and thinking allows you to ask tough questions and, as a result, you may ask the "Why" of anything; and, this sets the stage for a new project, never built before and never seen before.

Why wait until college to experience college? Why pay college fees to have the opportunity to use a $200,000 microscope? Why do elementary schools look the way they do? Why can't a student from a low-income family have a college-like education in an elementary school setting? The list of asking the "Why" of things was very extensive for this particular project and, as a result, the project goals and mission were defined. The project scope was so well documented, it organized the entire project even before procurement for the team took place. The project manager worked closely with decision makers and stakeholders to set the stage for the very first project ever to completely break the mold of a typical elementary school, and allow 8th and 9th grade students attend classes in a state-of-the-art school building, designed and furnished as a college building.

The challenges for this STEM (Science/Technology/Engineering/Mathematics) project were dictated by the clearly defined scope, budget, and schedule, in that exact order. The "Why" of things lead to ambitious dreams, which were explained as visions, and transferred into a detailed written document for the entire team to follow. From the very first time this project was verbally explained, the WOW factor began working its magic, and until this date, 2 years after being completed and now occupied by students, the WOW factor continues.

As a project manager, amazing opportunities will come and knock on your door. Although they may seem impossible to attain, and the fear of failure may be present, trust in your abilities and jump on board to the opportunity to manage a significant and meaningful project. When you do, you allow yourself to become a well-rounded project manager. Project management does open many doors so you may use your unique talents and abilities, and apply your acquired knowledge to projects. These types of projects often change an entire industry and break new ground to a more efficient and effective way of being and doing things.

Dream often of someday managing a project so ambitious that it scares you to the point that it becomes impossible to ignore it. The idea of being associated with such a project is constantly in your mind, calling you out. Say yes, and choose to look for those unique projects that make you better and allow you to grow. Take pride in being efficient and managing every dollar to be wisely spent to accomplish the main goals and aspirations of the project. Turn your dream into a vision, and take action as you begin managing projects with a new set of rules in mind. It is true; you must dream first before your dreams come true.

PM Benefits Your Business Profitability

Learning to delegate responsibilities is an ability that you, as a project manager, develop once you fully comprehend, through extensive experience, what people can do for you and the project. Project managers are experts in understanding what each team member is responsible for, and what each and every one is capable of contributing to the progress and completion of the project.

You, acting as a project manager, are the only individual in the entire organization who understands who everybody is, and who and what each stakeholder represents. Project owners, on the other hand, always delegate tasks to the inappropriate person or party. For example, they ask engineers questions meant for architects, and ask architects to address issues meant for a general contractor to address. The previous are just one of many from an extensive list of things some stakeholders may do wrong; and 99.9% of the time, their mistakes cost them money—a lot of money. Early on, in Chapter one, you read about the results for lack of proper project management. Inappropriately assigning tasks is the result of not knowing or not having a project manager assisting the project.

Therefore, you may confidently state that having you as a project manager, in responsible charge to manage the development of any project, benefits the profitability of the organization. Yes, of course, you are responsible for budget, and with that assumed duty, watching how every dollar is spent. Your knowledge and techniques to understand and verify costs are highly beneficial to any project owner. Polish your skills to review bids, quotes, and line item costs on a GMP (Guaranteed Maximum Price) document.

There are many techniques, sophisticated software, and experts out there to assist you in your efforts to make smart and realistic decision, as you verify costs are in line with the current economy or the industry standards. Professional fees are based on services provided, project budget, or a made up number; therefore, you must learn to recognize when each instance is presented to you. Polish your negotiation skills, and create partnerships with companies and vendors who may assist you in your understanding for all types of costs in project development.

Money allows the project to start. Then, it allows the project to move forward; and last, it allows for the completion of the project. Money is truly the heart of the project—without it, the project may not survive.

As you become more efficient and knowledgeable about reviewing and approving costs, your ability to reduce spending by the millions of dollars will increase. It will also allow your project management skills to benefit any organization financially.

 *To increase your business's bottom line, and maximize profit, please visit **www.thebookonpm.com**, and schedule a professional consultation with an experienced project manager.*

Chapter 10
How To Assess Your Results

Every Project is Uniquely Different

Even if you work in franchise project development, and every project has the exact same requirements and goals so they all look the same, the simple fact that the project is in a different location will make that project uniquely different. It is true: no two projects are alike, and as you recognize that every project has a unique set of rules and variables, you will begin to adapt to treat and properly manage the differences in your projects. In a way, the variables and differences in a project is what makes project management enjoyable. And truthfully, your ability to handle the differences in diverse projects is what makes you a unique project manager.

Project management expands across several industries, and is a profession practiced worldwide. Since scope for projects are diverse, procurement and team members are also diverse. As a result, anytime people and personalities, and their personal experiences, are part of the process of developing a project, the project becomes uniquely different. It is like a meal using the exact same ingredients but cooked by different chefs; the taste and even the way the meal is served on the plate will be different. Subsequently, it is a smart practice to systemize your processes used during project development to become more efficient, and to allow repetition of tasks, so the parts of your process become automatic and predictable.

There are valuable practices or tactics to allow constant growth of your project management knowledge and skills. First, learn to always see the entire project, from zero to infinity, meaning that you see it from conception to completion and beyond. As you visualize the entire process and live it in your mind, write down areas or phases that stand out, especially those phases or specific tasks where you fall short of the required experience; note all items that will require the participation of an expert. Second, have a robust procurement plan, and learn how to evaluate potential team members based on the actual project requirements they will be working on, and not so much on their previous successes, especially if the tasks they are handling this time are different than their past experiences.

Third, learn how to exercise accountability. By this, it is meant that you learn who is who in your project, and what their precise responsibilities are, so you are able to place those duties in a formal contract to hold them accountable. If some stakeholders are not part of any contract, create a document, such as an organizational chart, and outline their duties and scope of responsibilities. This document may be presented and approved by them, and that may become their contract. Lastly, commit yourself to always finding solutions as you manage the entire project. These values, along with all the new skills and knowledge you will gain, can become a set of tools you may take to any project you are managing, regardless of how uniquely different they may be.

How to Assess Your Results

A Measuring Stick

How do you measure your project's success, and how does a project make it on your list of successful projects?

Answers to these questions will vary depending on the types of project and the type of project manager you are. There are many indicators that tell you that your project is going well, and you are being successful. For instance, a happy owner who is recommending you to other clients, to the point that he already sold your services, and all that is left to do is to sign the contract. Perhaps success shows up as passing every inspection with flying colors, and you earn the respect of inspectors. For many people who trade their talent and time for money, the ultimate indicator comes via a significant bonus at the end of the year. So, you may have all or some of the previously mentioned indicators, but still, other important factors may not be present, such as the emotional wellbeing of the entire team—possibly feeling they had to work extra hours, and feel unvalued or unappreciated. If people in your team are staying away from you, and you never get invited to lunch or to have a meaningful conversation outside project topics, or you never pass time with your team, what kind of indicator would that be? Moreover, there may be vendors who may feel they were strategically obligated to do things outside their contract just to gain the opportunity to do business in future projects. One more, what about the politics of the project, and the fact that powerful stakeholders moved the project ahead and skipped the required process, leaving other stakeholders unappreciated or simply feeling ignored and not valued? When half of the indicators are not present, and you receive an award, how would you feel about being recognized? On the other hand, if all indicators are there and no award is given, you will most likely feel accomplished and deserving any and all credit and recognition.

It is nearly impossible to measure the success of a project, unless everything and everybody is completely satisfied and fully pleased. Having stated this, how about outlining, during the planning process and scope definition, what would be considered a success for the project you are managing? This may give you a clear target, easy to identify, and a set of guidelines and expectations regarding the positive outcome of the project. So, do it. Ask, and document a definition for success for your project in your detailed scope, vision, and mission, or your *goals documents*. Then, work to get there.

As an efficient and effective project manager, measuring the project beyond successful compliance with scope, budget, and schedule is a smart thing to do. Having a detailed description or definition for success in your project also allows you to realize when you go the extra mile, and receive the credit and recognition for doing so. Your very own personal measuring stick plays an important role in how you manage a project. Strive to compare your management style and practices with those of people you admire, respect, and consider mentors.

Defining your own measuring stick is healthy and required, so when your job is done and measured, you are able to say, "I have done my best and then some more." Ultimately, your longevity in project management will be mostly decided on how you manage all different measuring sticks, assessing your true value.

Close Out

It's time to finalize all project activities and verify that everything is entirely completed across all phases. As you enter this phase and close the project to transfer the completed project as required, the end of the road for the majority of your team

How to Assess Your Results

members is near. You, as the project manager, have some critical and important steps to take before you sign off on the closing of all phases. During this phase, it is critical to involve all project participants and stakeholders, and use a robust checklist to make sure you cover each and every item that has been completed.

> *When I guide your company's team to close out the project, I use clever techniques that will maximize the effectiveness of this phase. Please visit* **www.thebookonpm.com**, *and schedule a professional consultation to help your team during close out.*

Soliciting feedback as you conduct a post-project survey is more opportunity to understand how the entire team feels about the project as a whole, and in specific areas outlined in your survey. Including a *Lessons Learned* meeting allows you to gather useful information for the benefit and success of future projects. During this phase, the collecting of project data to be archived is the actual deadline for the majority of the team members. If your project requires archives to comply with any regulations, such as NARA (US National Archives and Records Administration), or others, verify that all submitted documents and data are in full compliance.

The checklist pertaining to your close out procedure may be very extensive as every project has its own unique requirements; stakeholders often require specific information, such as financial audits, or they may have confidentiality requirements and disposal of sensitive information, in order to properly close out a project. Ideally, your checklist for close out phase is amended with every project you manage, and it is your practice to always include a formal meeting with stakeholders prior to arriving at this particular phase.

Useful Life Costs

For many developments, project management ends at the completion of a project, once the owner takes full control of the completed product. Specifically, in building project development, once the certificate of occupancy is issued by the government entity, and the owner occupies the completed building, and there are no outstanding issues with punch list items or processing of the final payment, the project manager is typically ready to move to the next project. Infrequently, the project manager participates in managing the warranty period, typically lasting up to two years. If the project delivery method was via design-build-maintain, you, as the project manager, may be asked to remain in the project since you are the official holder of the information pertaining to just about everything regarding the completed project. If you are given the opportunity to participate in the management of a completed project, such as building while occupied by its end users, and get to see its operations, as well as experience the maintenance phase, your ability to define scope will drastically and beneficially improve. Furthermore, your ability to see the whole picture, and have a better watchful eye during production, will get sharper since you will have important knowledge about why things fail and need to be repaired during the two year occupancy period, in the case of a building.

In project development, knowledge is power; hence, if you have direct access to gain a deeper understanding regarding operating costs, maintenance, upgrade requirements, and replacement costs, and learn about the useful life of a product during the life cycle of a project, the better you will become at project delivery processes, addressing all aspects of planning and budgeting, specifically during design and construction. You will also be able to forecast both pending and future costs of several vital elements of any project in the areas of energy, mechanical, and electrical systems, as well as building envelope and structural systems.

As a project manager, you must make the project owner aware of the importance of performing a *life cycle cost analysis*, which simply means a complete understanding of total costs of ownership. Having the financial means to develop a project until the point of making it functional and operational is considered by most a healthy project budget.

Eventually, the daily operational costs requirements of the building begin, along with its demands for repairs and replacements, and keeping up with expenses becomes such a big challenge that business are closed, building are abandoned, or the functionality, as in the case of a building, is limited.

*To benefit from the services of an experienced project manager, please visit **www.thebookonpm.com**, and schedule a professional consultation.*

Next project, as you participate during review and approval of a budget, work with your team to include a *life cycle costs analysis*—one that is as real as if the building had already been built and been under operation for two years or more. Then, watch how priorities are shifted, and money is assigned to things that make the building better and as close to problem-free as possible to reduce the impact of operational and maintenance costs throughout the useful intended life of the project.

Sustainability

How long can anything be maintained at a certain rate or level? Even more specific, how long can a project sustain its healthy life while serving its intended function? And lastly, for how many years can it be sustained without undesirably impacting its surroundings or the environment?

The UN World Commission on Environment and Development's definition for sustainability reads: *"Sustainable development is development that meets the needs of the present without compromising the ability of future generations to meet their own needs."* Now, since buildings require money to stay alive, let us look at financial sustainability, and imagine two persons sitting side by side, each having one dollar. The only way these two people may remain side by side is to give a dollar, and then receive a dollar. So, person "A" gives a dollar to Person "B" at the same time person "B" gives a dollar to person "A". This process, as long as it happens precisely as described, can be sustained indefinitely. There may be multiple streams of income, including sophisticated systems for income generation, coupled with a sound administration for all finances; and your odds for remaining sustainable just increased.

During project management, absolutely everything you do affects the sustainability of a project in its broad definition. Therefore, sustainability must always be part of your big picture and original vision. It is irresponsible to develop a project without considering life cycle and its sustainability. Efficiency, for many project managers, only means to work less and accomplish the same or more. For the financial managers, it means to pay less but build the same or more. Those are very limited ways of thinking, and your efficiency definition is to be inclusive of sustainability always accounting for the life expectancy of a project.

Imagine you are managing a project, and you implemented life cycle costs as part of the budget, and complemented that by a full sustainability analysis for all three functional, environmental, and financial reasons. The life expectancy of the project or building is 50 years. Fast forward to 50 years from today: you receive a phone call, an email, or a letter, notifying you that you are receiving an award as part of a team who planned and implemented a perfect plan for the building to be sustained fully operational throughout its life, without the need to compromise anything or be in financial hardship. The name of the award is, *The A and B Infinite Dollar Award.*

Every project you manage is to include the commonly avoided sustainability conversation, since many project owners usually ignore it. Carefully bring this subject up, and address its requirements. Working closely with all stakeholders, do your best to utilize the entire team's abilities. If you must, hire expert advice. Perhaps and in 30 years or less, you may check your email or your PO BOX and an award notice on "Exceptional Sustainability" will be there as one more proof of your commitment and success in accomplishing a sustainable project.

Decode, Decrypt, and Decipher All About It

Documenting the entire process, and creating a reliable system to access it, read it, and understand it, may be one of the major challenges the project development industry faces. With each project that gets developed, you, as the project manager, are the most knowledgeable of any other entity in the organization or development team. The information produced before, during, and after a project in development is vast, as it is inclusive of everything in relation to the entire life cycle of that project.

The main purpose for documenting a project is to have the ability to understand how, what, who, when, and why things were developed a certain way. There are many processes to organize, archive, and access information; however, no software out in the market does it all. Hence, let us list a few important steps to consider while you make a decision on how to collect and store project data for future use.

The steps below are guidelines to consider as you come up with your own unique methodology:

1. Whenever possible, collect all data in digital form.
2. Use a file naming system that is descriptive; 30 years from today, a stranger can decipher what is in that file. If names are too long, create a file under each category, with a key or an index of how files were organized.
3. Always document information for a stranger to read it and understand it. Avoid documenting things in a way only you or team members can decipher.
4. When saving files, use an extension that has been widely used for more than 10 years, and an extension that is easily accepted and used by 3 generations, with a year span of 20 years in between. For example, PDF is the most common worldwide accepted format for digital files, and it first arrived in June 1993. This extension has been used for over 24 years, and people from 13 to 63 years of age are familiar with it and can use it.
5. Make sure software to read the digital files is free and does not cost the viewer anything to install. Arrange files under descriptive folders, and locate folders under each category. Include all pertinent communications under each of the project phases.
6. Write instructions at the beginning of the data, explaining how you have organized the project information, and keep copies of the entire database in a remote server. Give digital copies to key stakeholders and, if possible, create an instructional video on how to access, find, and decipher all about the project you managed, with them in mind.

How to Assess Your Results

Follow the above listed steps and, 30 years from today, anybody who accesses the project database will be able to fully understand as they decode, decrypt, and decipher all about your project.

> *When I train teams or individuals on the effective ways to collect data, I use unique methodologies to eliminate the need for guessing any project information. Please visit **www.thebookonpm.com**, and schedule a consultation to assist you in perfecting your data collecting processes.*

A Final Note

Now, give yourself credit. Your project management knowledge, skills, and tools have been complemented just by the mere fact you have read this book. Your journey as a project manager will be filled with amazing opportunities to leave this world a better place. Your participation in project management, along with millions of fellow project managers around the globe, will add to the efforts of the fastest growing profession in the world. Be efficient and honest, and stay true to your values; surprise yourself first, and every project you manage will become successful in every way, exactly as you imagined it.

Share your story; spread the joy and your knowledge to those around you. You are already an amazing project manager. All that is left to do is for you to let it come out.

It is time to play! Enjoy your PM journey.

About the Author

Gustavo A. Valenzuela was born near the beautiful beaches of the Sea of Cortez, in Guaymas, Sonora, Mexico, where the Sonoran Desert meets the blue water of the Gulf of California, in the Pacific Ocean. At age 17, in August of 1987, he moved to the United States of America to attend his senior high school year at Liberty Union High School, in Baltimore, Ohio, where he lived as an exchange student with The Talbots family. He found the USA to be his home as he has happily and productively lived in this country since then.

In 1988, Gustavo attended Pima Community College, where he tuned up his English speaking and writing skills, while he astutely immersed himself in advanced construction classes. He lived in Tucson, Arizona, while attending college, and worked full time building homes from the ground up, where he could learn every construction trade. The company he worked for at that time delivered projects via turn-key, which means every project was developed from conception until completion, and literally delivered the building keys to the owner, once fully functional.

Gustavo continued to advance his life; he applied and was accepted to the College of Architecture, at the University of Arizona. He graduated with a Bachelor of Architecture, in 1997, and a few years later, he became a licensed Commercial/

Residential General Contractor. For the past 25 years, Gustavo has dedicated his life to becoming an expert in architecture, engineering, and construction. During this lapse of time, he designed, built, and successfully managed a myriad of projects in the United States, ranging in budget from $50,000 to over 22 million dollars. He firmly believes that becoming an expert in all that you take on is an ability and a commitment, which he encourages you to also incorporate as part of your life.

Gustavo embraces all his core values, especially his continuous learning and reliability values; therefore, on April 24, 2016, he was certified as a life coach to complement his tools, and enhance the experience of all the people he can support. Each day, he reminds himself to access his toolbox and use as many tools as possible to live each day with meaning and purpose.

Gustavo also works on elevating his vision. He is constantly planning, designing, and visualizing his next life project; his vision remains clear, powerful, and meaningful. In his life schedule, he looks forward to important tasks being completed. Therefore, he is committedly working to run a worldwide fundraiser to raise 33 million dollars to build a spiritual center. He visualizes this project as an opportunity for people from all over the world to come and visit this unique center, to permanently heal emotionally and spiritually, as they interact with the unique architecture and the healing energy of the entire complex. As one gets to know Gustavo, one realizes that he is simply grateful for being able to effortlessly find new meaning, and connect all things with a higher purpose. Consequently, his life's vision has encountered new meaning. He faithfully and clearly visualizes himself alongside 32 other known and unknown leaders, coming together to combine all their knowledge, and utilize all their tools acquired during their life's

About the Author

journey, to build a unique, artistic, and inspiring architectural piece to help anybody find a kind of inner peace seldom experienced. Gustavo's spiritual healing complex will provide a space and an opportunity for every visitor to immerse themselves, in order to naturally activate forgiveness, release any blocks, and blissfully open to connect with their authentic being, and continue to live a positive and meaningful life.

Gustavo's ability to share his big dreams is typical of his personality, and is clearly printed in his innate nature. He respectfully shares the information in this book with you, with high hopes that he can inspire you to elevate your project management skills as you begin or continue your journey as a project manager.

Gustavo, as the author of this book, truly aspires to make a significant impact and difference in your professional and personal life as a project manager. He is available to help you along your journey, and he is enthusiastic to answer any questions you may have regarding any of the topics covered in this book.

Gustavo may be contacted by email, at the address listed below, or at **gustavo@the bookonpm.com.**

Gustavo A. Valenzuela
PO BOX 38100
Phoenix Arizona 85069 USA

If you would like to contact Gustavo for seminars or specialty trainings on project management, or you would like Gustavo to become part of your team of experts in a project, please email him or contact him via the website at **www.thebookonpm.com.**

www.ingramcontent.com/pod-product-compliance
Lightning Source LLC
Chambersburg PA
CBHW070145230526
45471CB00002B/524